Making Geometry Come Alive

Student Activities & Teacher Notes

ALFRED S. POSAMENTIER

For information:

Corwin Press, Inc.
A Sage Publications Company
2455 Teller Road
Thousand Oaks, California 91320
E-mail: order@corwinpress.com

Sage Publications Ltd.
6 Bonhill Street
London EC2A 4PU
United Kingdom

Sage Publications India Pvt. Ltd.
M-32 Market
Greater Kailash I
New Delhi 110 048 India

Printed in the United States of America

Library of Congress Cataloging-in-Publication Data

Posamentier, Alfred S.
Making geometry come alive: Student activities and teacher notes / by Alfred S. Posamentier.
p. cm. — (Math assessment series)
ISBN 0-7619-7598-5 (c) — ISBN 0-7619-7599-3 (p)
1. Geometry—Problems, exercises, etc. I. Title. II. Series.
QA459 .P69 2000
516—dc21 00-008375

This book is printed on acid-free paper.

00 01 02 03 04 05 10 9 8 7 6 5 4 3 2 1

Editorial Assistant: Catherine Kantor
Production Editor: Diana E. Axelsen
Editorial Assistant: Victoria Cheng
Typesetter/Designer: Technical Typesetting, Inc.
Designer: Tracy Miller

CONTENTS

Introduction

Making Geometry Come Alive is a set of versatile enrichment exercises that cover a very broad range of topics in geometry—Euclidean, post-Euclidean, and non-Euclidean. Several criteria have been used to develop the activities and to select the topics that are included. All of them bear heavily, and equally, on my concerns for curriculum goals and classroom management.

First and foremost, the activities are meant to be motivational. As much as possible, I want this book to achieve the goal of being attractive to students and to show them that there is much more to geometry than proving theorems. To demonstrate this aspect of geometry, it is necessary for the activity to be quite different from what students encounter in their basal texts—different in both substance and form. This seems especially critical; no matter how excellently a basal text is being used, nearly every class experiences the "blahs." Unfortunately, this sort of boredom is often well entrenched long before the teacher and perhaps even the students are aware of it. Presenting activities on a regular basis gives the variety and change of pace needed to sustain interest in any subject.

With the large number of topics you may have to cover during the normal school year, it may seem naïve or unrealistic to suggest introducing additional material. This brings me to the second criterion. Most of the activities in this volume can be used to enhance, reinforce, and extend the concepts and skills that already make up the better part of your curriculum and course goals.

An inspection of the first three sections of this book quickly reveals how this is so. These 15 activities draw upon the full arsenal of what you've given your students from the theorems, postulates, and exercises in their basal texts. The activities herein are designed as an aid to presenting the basic concepts of your course, as well as a set of motivational and enrichment activities. These objectives are completely consistent with the *Principles and Standards for School Mathematics* (NCTM, 2000). These activities offer a good practice in what you're trying to teach anyway and also greatly increase your students' awareness of the different directions in which these ideas can lead.

A third criterion, one special to this volume, is the linkage between algebra and geometry. Euclid, Pythagoras, and their contemporaries were unaware of algebra and, unfortunately, many teachers through many centuries have taught geometry as though algebra didn't exist. Where possible in this book, I have tried to show this important connection.

Another criterion was that each activity should have some use or merit beyond itself, a heuristic value. That is, the activities serve as door openers—introductions to areas not usually treated in basal texts. The non-Euclidean geometry units are good examples of this.

The Key: Problem Solving

Finally, the activities provide opportunities and incentives to hone problem-solving skills—not merely chapter-end exercises that are *often called* problems, but realistic problems such as your students will encounter in their everyday living and in later, nonmathematics school courses. Most of the activities begin by posing a problem that students find intriguing and that, at the outset, many students are unable to solve on their own. In working through the problem, however, the students discover they can tackle a much bigger monster than they thought they were capable of handling. Equally important, they find these problem-solving techniques are applicable to other areas.

The problem-solving orientation of these activities cannot be overemphasized. Those of you who have read the NCTM Standards (2000) probably agree that its recommendations, however difficult to implement, are right on target. The *Math Alive Series* is a deliberate step toward achieving these goals.

Presenting the Activities

In pilot testing these activities, I worked with teachers who had very diverse mathematical preparation and who had to deal with a *wide* spectrum of class size, student ability, and class heterogeneity. Thus, it seemed very desirable to search for alternative means to present the activities. I discovered several. One or more of them should be useful in your situation.

The normal presentation, the one that best suits most classes, is to present the activity as a new lesson at the outset of a class period. In working through the student page, you'll find the accompanying Teacher's Notes explain the rationale for the entire activity, as well as provide anticipated student responses and questions. However familiar you feel with the mathematical topics presented, do not attempt to conduct a class session without first spending 20 or 30 minutes going over the Teacher's Notes. Both the student pages and the Teacher's Notes are highly compressed: A typical student page encompasses the concepts that four or five basal-text pages generally treat.

In some cases, the student activity can be handed out the day preceding class discussion. Your perusal of the activity will best determine when this is appropriate. In many other cases, you will find it best to discuss only the body of the student activity the day you pass it out, deferring the discussion of the Extension until the following day.

If your class is like many that I have encountered, you may wish to try peer teaching. This has many advantages for both you and your students if your classes have three to six really bright students. By giving both the student page and Teacher's Notes to one of these "stars," he or she can present the activity the following day to this group of above-average students. This allows you time to work with your average and below-average students to bring their skills up to par without boring the students who are already well on top of things. My experience has shown that students who are asked to present activities prepare very well. Their pride is at stake, and thus you can be sure they won't let you down.

The Extensions

The Extensions offer the greatest opportunity for flexibility in using the activities. Every activity in these volumes has one, but they differ. In some cases, they dip into more sophisticated mathematical concepts and should be considered as optional activities primarily for your better students. In other cases, the Extensions require no additional mathematical sophistication, but simply give an opportunity to explore the topic in greater detail. Your reading of the activity will quickly determine which is the case. Sometimes you may want to present the basic activity to the class and assign the Extension as homework for your better students. In all cases, you should think of the Extension as an element that allows you to tailor your mathematics program to best meet the needs and interests of all your students.

Selecting the Activities by Topic

This volume probably contains more activities than you'll be able to use in a single school year. The chapter introductions will assist you in selecting the activities best suited to your students' abilities and interests and offer some hints as to how they can be used. The activities have been divided into seven categories. The difficulty level among the activities varies, so your assessment with respect to your classes' interests and abilities is of paramount importance.

The following pages give an overview of the activities, category by category. A diamond (♦) follows the titles of the activities that are in reach of your slower students. A star (⋆) indicates the activities that are probably best given only to your better students or given a more careful presentation to the general class. (Note that a ♦ doesn't mean that your better students won't like the activity it simply means the activity is within the grasp of your mathematically less proficient ones.)

The NCTM Principles and Standards for School Mathematics – 2000

Each unit is tied in with one or more of the NCTM Standards presented in *Principles and Standards for School Mathematics* – 2000. As units are selected for use in the classroom, it is good to be aware of the Standards being employed. A simple numbering system is used to help make this identification simple and unobtrusive. At the start of each "Teacher Notes" section, the Standard number appropriate for that unit is indicated by a dot below the appropriate Standard number. These numbers correspond to the following list of standards:

1. Number and Operations Standard

Instructional programs from prekindergarten through grade 12 should enable all students to –

- understand numbers, ways of representing numbers, relationships among numbers, and number systems;
- understand meanings of operations and how they relate to one another;
- compute fluently and make reasonable estimates.

2. Algebra Standard

Instructional programs from prekindergarten through grade 12 should enable all students to –

- understand patterns, relations, and functions;
- represent and analyze mathematical situations and structures using algebraic symbols;
- use mathematical models to represent and understand quantitative relationships;
- analyze change in various contexts.

3. Geometry Standard

Instructional programs from prekindergarten through grade 12 should enable all students to –

- analyze characteristics and properties of two- and three-dimensional geometric shapes and develop mathematical arguments about geometric relationships;
- specify location and describe spatial relationships using coordinate geometry and other representational systems;
- apply transformations and use symmetry to analyze mathematical situations;
- use visualization, spatial reasoning, and geometric modeling to solve problems.

4. Measurement Standard

Instructional programs from prekindergarten through grade 12 should enable all students to –

- understand measurable attributes of objects and the units, systems, and processes of measurement;
- apply appropriate techniques, tools, and formulas to determine measurements.

5. Data Analysis and Probability Standard

Instructional programs from prekindergarten through grade 12 should enable all students to –

- formulate questions that can be addressed with data and collect, organize, and display relevant data to answer them;
- select and use appropriate statistical methods to analyze data;
- develop and evaluate inferences and predictions that are based on data;
- understand and apply basic concepts of probability.

6. Problem Solving Standard

Instructional programs from prekindergarten through grade 12 should enable all students to –

- build new mathematical knowledge through problem solving;
- solve problems that arise in mathematics and in other contexts;
- apply and adapt a wide variety of appropriate strategies to solve problems;
- monitor and reflect on the process of mathematical problem solving.

7. Reasoning and Proof Standard

Instructional programs from prekindergarten through grade 12 should enable all students to –

- recognize reasoning and proof as fundamental aspects of mathematics;
- make and investigate mathematical conjectures;
- develop and evaluate mathematical arguments and proofs;
- select and use various types of reasoning and methods of proof.

8. Communication Standard

Instructional programs from prekindergarten through grade 12 should enable all students to –

- organize and consolidate their mathematical thinking through communication;
- communicate their mathematical thinking coherently and clearly to peers, teachers, and others;
- analyze and evaluate the mathematical thinking and strategies of others;
- use the language of mathematics to express mathematical ideas precisely.

9. Connections Standard

Instructional programs from prekindergarten through grade 12 should enable all students to –

- recognize and use connections among different mathematical ideas;
- understand how mathematical ideas interconnect and build one another to produce a coherent whole;
- recognize and apply mathematics in contexts outside of mathematics.

10. Representation Standard

Instructional programs from prekindergarten through grade 12 should enable all students to –

- create and use representations to organize, record, and communicate mathematical ideas;
- select; apply, and translate among mathematical representations to solve problems;
- use representations to model and interpret physical, social, and mathematical phenomena.

ABOUT THE AUTHOR

Alfred S. Posamentier is Professor of Mathematics Education and Dean of the School of Education of the City College of the City University of New York. He is the author and coauthor of numerous mathematics books for teachers and secondary school students. As a guest lecturer, he favors topics regarding aspects of mathematics problem solving and the introduction of uncommon topics into the secondary school realm for the purpose of enriching the mathematics experience of those students. The development of this book reflects these penchants.

After completing his A.B. degree in mathematics at Hunter College of the City University of New York, he took a position as a teacher of mathematics at Theodore Roosevelt High School in the Bronx (New York), where he focused his attention on improving the students' problem-solving skills. He also developed the school's first mathematics teams (at both the junior and senior level) and established a special class whose primary focus was on mathematics problem solving and enrichment topics in mathematics.

For years, Dr. Posamentier has collected clever ways of introducing students to new concepts in mathematics. This collection of ideas prompted the development of this book. He is currently involved in working with mathematics teachers, locally and internationally, to help them better understand problem-solving strategies and alternative instructional strategies, so that they can comfortably incorporate them into their regular instructional program.

Immediately upon joining the faculty of the City College (after having received his masters' degree there), he began to develop inservice courses for secondary school mathematics teachers, including such special areas as recreational mathematics, problem solving in mathematics, and instructional alternatives for the classroom.

Dr. Posamentier received his Ph.D. from Fordham University (New York) in mathematics education and has since extended his reputation in mathematics education to Europe. He is an Honorary Fellow at the South Bank University (London, England). He has been visiting professor at several Austrian, British, and German universities, most recently at the University of Vienna and at the Technical University of Vienna. At the former, he was a Fulbright Professor in 1990.

In recognition of his outstanding teaching, the City College Alumni Association named him Educator of the Year and had a day (May 1, 1993) named in his honor by the City Council President of New York City. In 1994, he was awarded the National Medal of Honor from the Austrian government, and in 1999, by an act of the Austrian Parliament he was awarded an Austrian University Professor title by the President of the Republic of Austria.

Naturally, with his penchant for motivating students towards mathematics, he has been very concerned that students have a proper introduction to mathematics from an entertaining point of view. This interest motivated the development of this book.

CHAPTER 1

Constructions

- Constructing Segments ♦
- Constructing Radical Lengths ♦
- Trisecting a Circle
- Trisecting an Angle
- Constructing a Pentagon
- Constructing Triangles ⋆

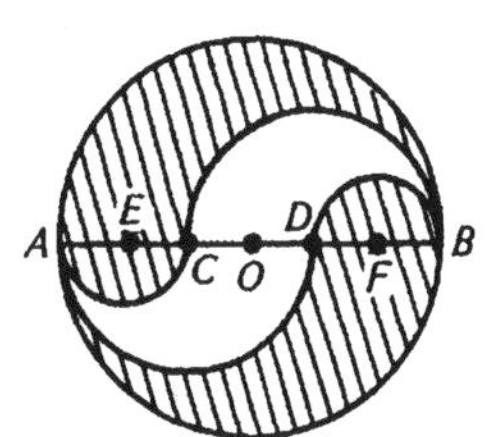

Constructions using a straightedge and compass are a basic part of any Euclidean geometry course. Unfortunately, many of the more interesting constructions are not explored. Neither are constructions presented in their most useful way, as reinforcement of many different geometric concepts and relationships and for the development of problem-solving skills.

The six activities in this category present specific construction problems, none of which is usually presented in high school geometry texts. These constructions are all proved either on the student page or in the Teacher's Notes.

One of the common ways to link algebra and geometry is through numerical applications of geometric theorems. "Constructing Segments" and "Constructing Radical Lengths" show how to construct lengths that represent sums, differences, products, quotients, and square roots of given lengths, given only the unit line segment. After these five operations are represented geometrically, students use them to represent complex algebraic expressions. The constructions may seem strange to students at first, so it is useful to assign numerical values to the variables. By making the constructions on graph paper, students can compare the algebraic and geometric answers.

"Trisecting a Circle" uses the area formula for a circle and the constructions developed in the previous two activities. Thus, it is best presented immediately after these activities; if not, the activities should be briefly reviewed. The three trisection methods that are presented result in three very different figures. Thus, students can see that unlike algebra, where there is usually a single right answer, there can be several correct answers to a geometry problem. The Extension is especially intriguing and should not be difficult for average students.

It is well known that the general angle cannot be trisected with Euclidean tools, and although "construction" usually implies using only a compass and a straightedge, there are other construction tools. "Trisecting an Angle" presents two methods of trisecting an angle using these other construction tools.

The method usually given for constructing a pentagon is both tedious and difficult to understand. The method presented in "Constructing a Pentagon" is easy to do *and* easy to justify. The justification is given in the Extension and only requires knowledge of similar triangles. This activity is closely related to "The Golden Rectangle" and "The Golden Triangle."

In the usual geometry course, students construct triangles using the congruence theorems, and most texts provide ample illustrations of these constructions. In "Constructing Triangles," three parts of a triangle are given and students are to construct the triangle from which they are taken. However, these parts include combinations of sides, altitudes, and medians rather than sides and angles. Only three cases are considered on the student page, but there are many other possibilities using the above-mentioned parts, as well as angles, angle bisectors, the semiperimeter, and the inradii and circumradii of a given triangle.

These triangle constructions not only review many important geometric concepts, but also help develop students' problem-solving skills. At first, students will need to see complete analyses of the constructions, but in time they will get the knack and begin to solve problems of increasing difficulty. "Constructing Triangles" is best presented near the end of the year so students can draw on more geometric concepts and relationships.

Once the facility and limitations of the straightedge and compasses are established, computer drawing programs, such as Geometer's Sketchpad, can be used.

Constructing Segments

Using a straightedge and compass, you can construct the sum, difference, product, and quotient of two given segments. That is, from segments a and b, you can construct $a + b$, $a - b$, ab, and $\frac{a}{b}$. This means that algebraic expressions can be represented geometrically. Constructing $a + b$ and $a - b$ is easy. Use the following segments with lengths a and b to construct a segment of length $a + b$ and a segment of length $a - b$.

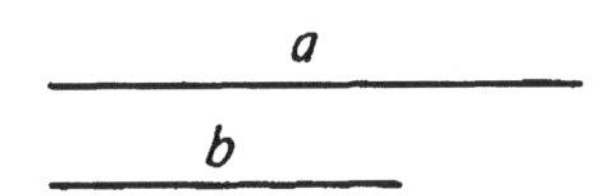

Constructing a segment of length ab is a little more difficult. In this case we need a unit segment in addition to our given segments. The unit segment and segments with lengths a and b are

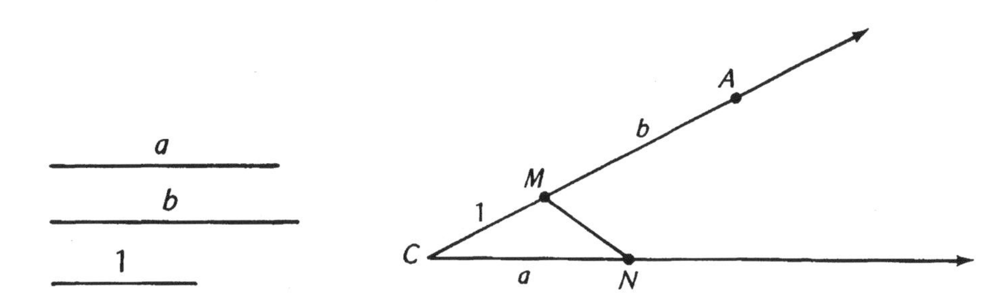

Begin by drawing any angle C and marking off the segments as show. Construct a line through A parallel to $\overline{MN}$ and intersecting $\overrightarrow{CN}$ at B. Label the length of $\overline{NB}$, x. Now consider the proportion $\frac{b}{1} = \frac{x}{a}$. Why is this proportion true?_____

____

What does x equal?________________________________

Notice that the unit segment is less than both a and b. That is, $a > 1$ and $b > 1$.

How should the product ab compare to a and b?____________________

Is this true in the preceding construction?________________________

Suppose $a < 1$ and $b < 1$ as in the segments

a

b

1

How should the product ab compare to a and b? ____________

Construct ab using the preceding method and $\angle C$:

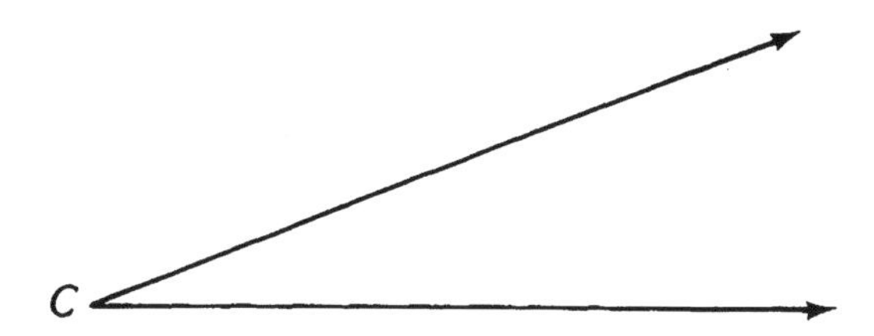

Does your construction verify your comparison of ab to a and b? ____________

We use a similar method to construct a segment of length $\frac{a}{b}$.

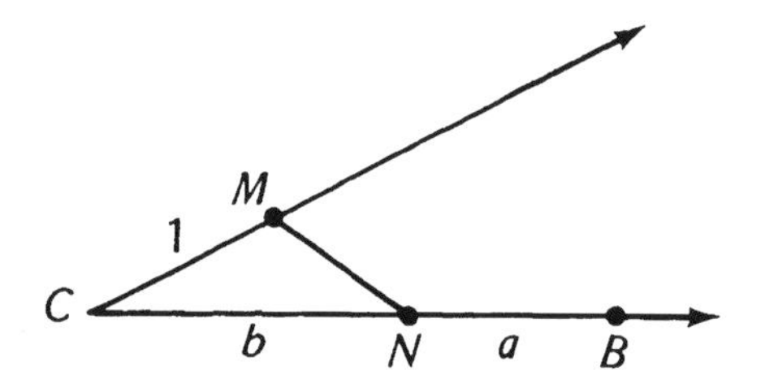

How can you locate A on $\overrightarrow{CM}$ so that $AM = \frac{a}{b}$?____________________

____

What proportion justifies this construction?____________________

Complete the construction.

EXTENSION! Given segments of length a, b, and c,

a

b

c

decide on a unit segment and construct $\frac{ab+c}{c-a}$. Now assign specific numerical values to a, b, and c to find the value of $\frac{ab+c}{c-a}$. Make your construction on graph paper and see if the answer you get geometrically agrees with your algebraic answer.

Teacher's Notes for Constructing Segments

Geometry students know how to use straightedge and compasses to copy segments and bisect segments. This activity shows them how to construct segments that represent algebraic expressions. Students will be surprised at how easy it is to construct $a+b$, $a-b$, ab, and $\frac{a}{b}$ given segments of lengths a and b. After students know these four constructions, they will be able to easily represent complicated algebraic expressions geometrically.

This activity should be presented immediately before "Constructing Radical Lengths," and after students have studied proportional segments in triangles.

NCTM Standards

1	2	3	4	5	6	7	8	9	10
	•	•	•		•	•		•	

Presenting the Activity

Students should be able to construct $a+b$ and $a-b$ without any additional instructions. They simply copy the segments on a straight line as shown:

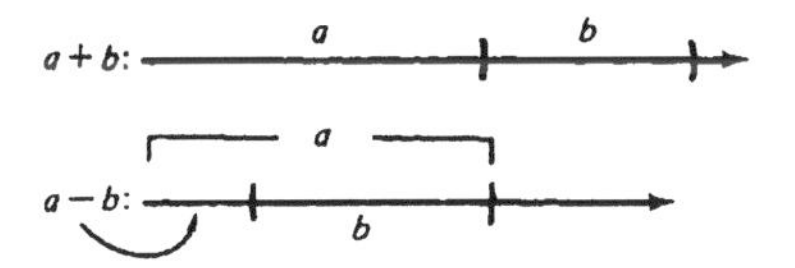

This, of course, is the same as adding and subtracting on a number line.

Students should also be able to easily complete the construction of ab. The proportion is true because a line parallel to one side of a triangle divides the other two sides proportionally. Solving the proportion for x gives $x = ab$. Thus, students have constructed the product of the lengths of the two given segments.

Because both $a > 1$ and $b > 1$, $ab > b$ and $ab > a$. It is easy to prove this algebraically: Multiply $a > 1$ by b and $b > 1$ by a. This relationship is true in the construction because the constructed segment of length ab is longer than both of the original segments.

For $a < 1$ and $b < 1$, $ab < b$ and $ab < a$. The following construction verifies this:

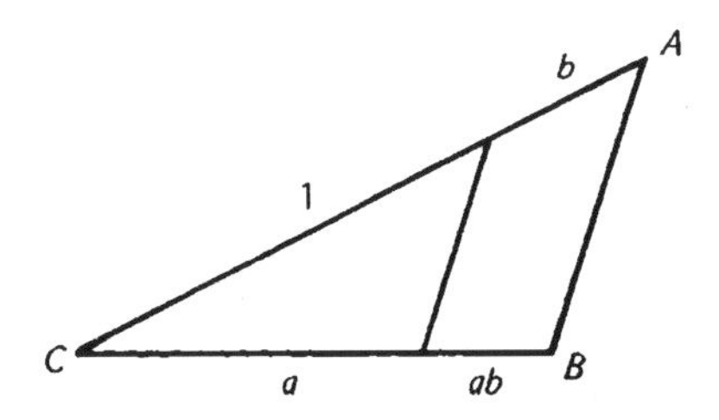

A similar method is used to construct $\frac{a}{b}$. Students should construct a line through B parallel to $\overline{MN}$ and intersecting $\overrightarrow{CM}$ at A. The proportion $\frac{a}{b} = \frac{x}{1}$, where x is the length of $\overline{MA}$, justifies the construction. The completed constructed is

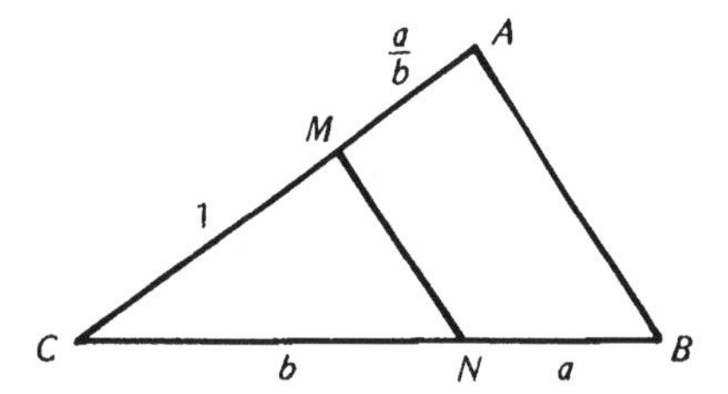

You may wish to have students explore the results for different lengths of a, b, and the unit segment. When $1 < a < b$, either $a < \frac{a}{b} < b < 1$ or $a < b < \frac{a}{b} < 1$; when $a < b < 1$, $b < a < \frac{a}{b}$. When $1 < a < b$, $\frac{a}{b} < 1 < a < b$ (this is the case given on the student page); and when $1 < b < a$, either $1 < \frac{a}{b} < b < a$ or $1 < b < \frac{a}{b} < a$.

Extension

Most students should be able to do this construction without much trouble. They simply work in steps, doing one construction at a time; as follows

Step 1. Construct ab:

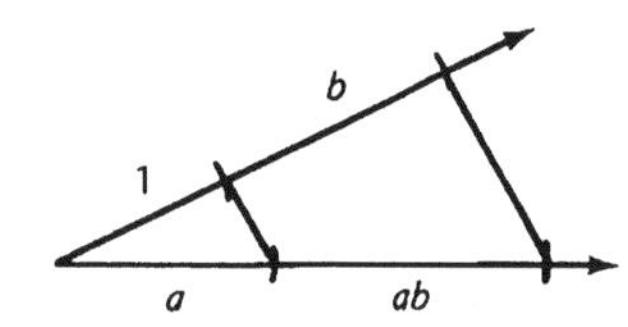

Step 2. Construct $ab + c$:

Step 3. Construct $c - a$:

Step 4. Construct $\frac{ab+c}{c-a}$:

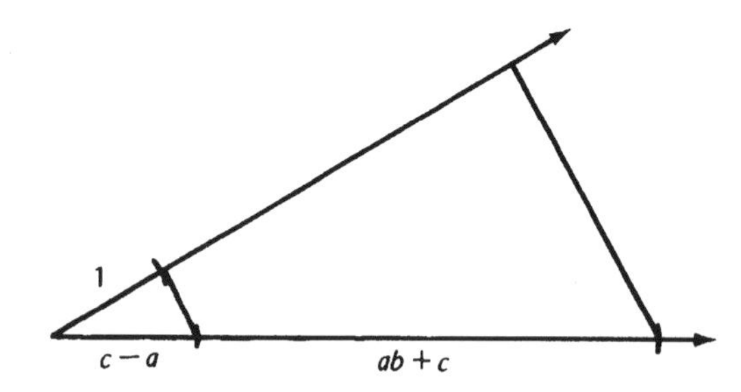

Students must decide on the length of the unit segment and should be reminded to use the same unit segment throughout. Note that different unit segments will produce different constructions.

Students may wish to explore constructing similar algebraic expressions. Point out that each element of the construction must be constructible. For example, using the techniques presented here, it isn't possible to construct $c - a$ if $a > c$ or if $c - a = 0$. Your advanced students may want to explore this.

Constructing Radical Lengths

In "Constructing Segments," you learned how to use a straightedge and a pair of compasses to construct $a + b$, $a - b$, ab, and $\frac{a}{b}$. It's also possible to construct a segment of length $\sqrt{a}$.

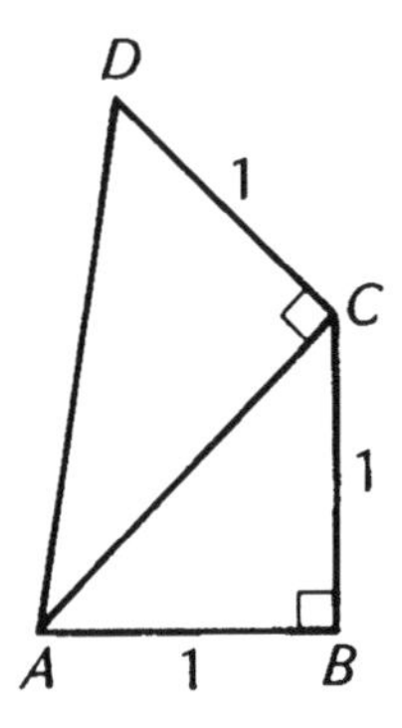

First, let's construct a segment of length $\sqrt{2}$. In isosceles right triangle ABC in the preceding diagram, how long is $\overline{AC}$, the hypotenuse?________________
How would you construct $\triangle ABC$?________________________________

____

In right $\triangle ACD$, how long is $\overline{AD}$, the hypotenuse? ____________ How would you construct $\triangle ACD$?____

____

Continue constructing right triangles using the hypotenuse of the preceding triangle as one leg and the unit length as the other leg until you have a segment of length $\sqrt{15}$. This figure is called a *radical spiral.*

The radical spiral gives us a method for constructing $\sqrt{a}$ for specific values of a. However, suppose we don't know the value of a? Then we use a unit segment and a different method to construct $\sqrt{a}$. A segment of length a and a unit segment are given as

a

1

Using the following line, start at A and mark off $\overline{AB}$ of length 1:

From B mark off $\overline{BC}$ of length a on the same line, such that $AC = 1 + a$. Construct the midpoint of $\overline{AC}$ and draw a semicircle that goes through points A

and C. Then construct a perpendicular at B intersecting the circle at D. The length of $\overline{BD}$ is $\sqrt{a}$.

This method can be used instead of the radical spiral to construct $\sqrt{a}$ for a specific value of a. On a separate sheet of paper, use the unit segment

1

to construct a segment of length $\sqrt{7}$.

EXTENSION! Given segments of length a, b, and c,

a

b

c

decide on a unit segment and construct $a\sqrt{\frac{2b-c}{a+c}}$.

Teacher's Notes for Constructing Radical Lengths

This activity should be presented immediately after "Constructing Segments," because it completes the geometric representation of algebraic operations. The constructions presented in these two activities will be used in "Trisecting a Circle." Remember, these constructions can also be done on a computer screen!

NCTM Standards

1	2	3	4	5	6	7	8	9	10
•	•	•	•		•	•	•	•	•

Presenting the Activity

Begin with a brief review of "Constructing Segments," pointing out that the only other algebraic operation for which a geometric representation is needed is finding square roots. Then have students consider isosceles right triangle ABC. By the Pythagorean theorem, $AC = \sqrt{2}$. To construct $\triangle ABC$, a perpendicular is constructed at a point B on a line. Then a unit segment is marked off at points A and C on the base line and perpendicular to the base line, respectively. Then $\overline{AC}$ is drawn.

Again using the Pythagorean theorem, the length of $\overline{AD}$ in $\triangle ACD$ is $\sqrt{3}$. Triangle ACD is constructed by first constructing a perpendicular to $\overline{AC}$ at C. Then $\overline{CD}$ of length 1 is marked off on the perpendicular and $\overline{AD}$ is drawn. The radical spiral to a segment of length $\sqrt{15}$ is

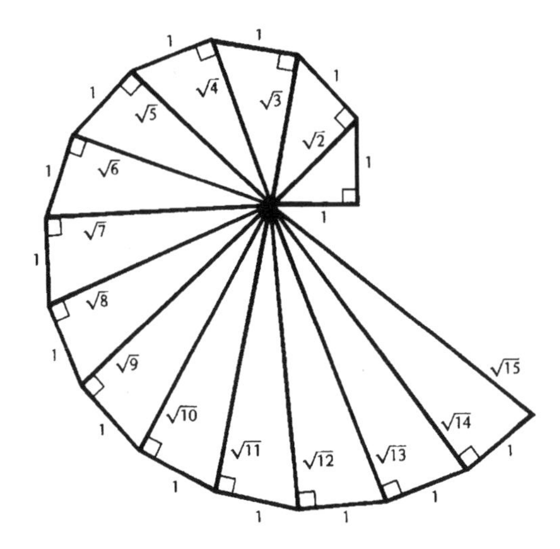

The next construction on the student page is

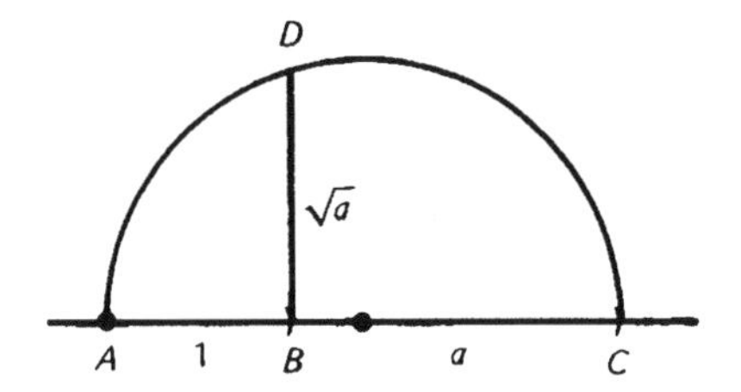

This construction can be justified as follows: By drawing $\overline{AD}$ and $\overline{DC}$, right $\triangle ADC$ is formed, because any angle inscribed in a semicircle is a right angle. The altitude to the hypotenuse of a right triangle is the *mean proportional* between the segments into which it divides the hypotenuse. $\overline{DB}$ is the altitude to the hypotenuse in $\triangle ADC$. Thus,

$$\frac{AB}{DB} = \frac{DB}{BC} \quad \text{or} \quad \frac{1}{DB} = \frac{DB}{a}.$$

Then $(DB)^2 = a$ and $DB = \sqrt{a}$. The construction of $\sqrt{7}$ using this method is

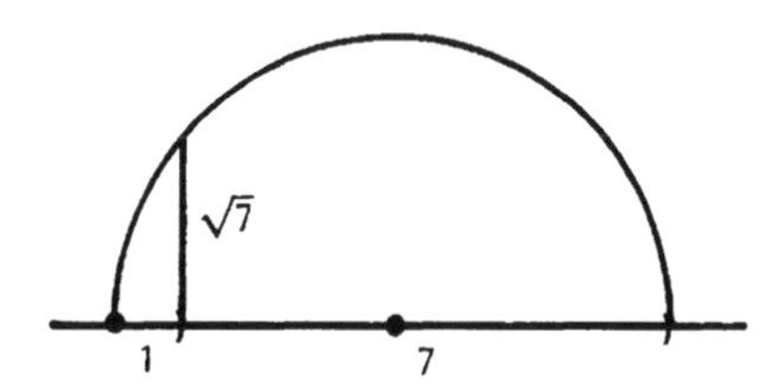

Extension

As in the Extension for "Constructing Segments," students should do one construction at a time. They must decide on a unit segment and should be reminded to use the same unit segment throughout.

Step 1. Construct $2b - c$:

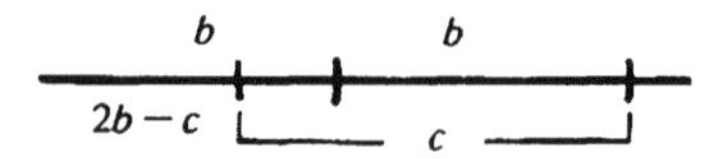

Step 2. Construct $a + c$:

Step 3. Construct $\frac{2b-c}{a+c}$:

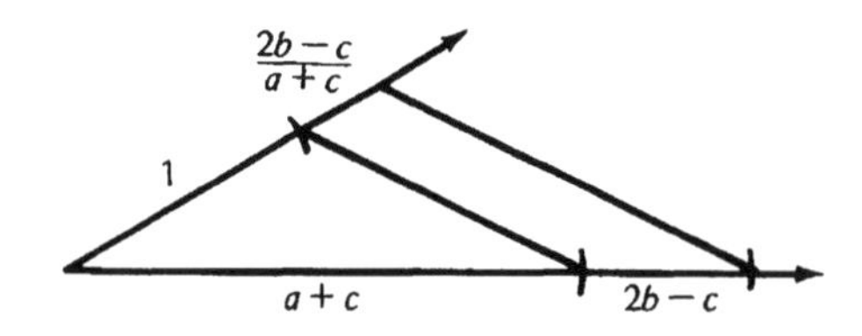

Step 4. Construct $\sqrt{\frac{2b-c}{a+c}}$:

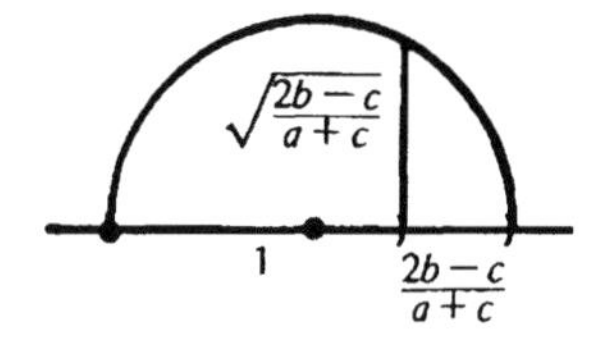

Step 5. Construct $a\sqrt{\frac{2b-c}{a+c}}$:

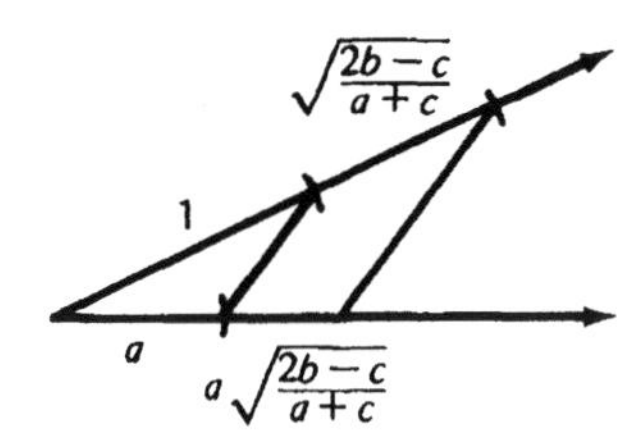

Trisecting a Circle

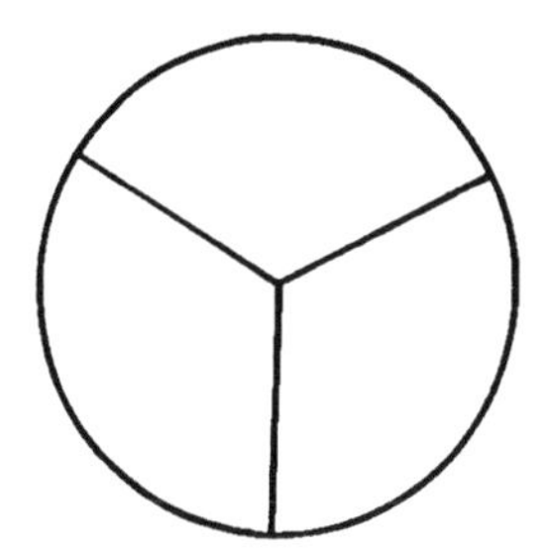

The preceding circle is divided into three regions of equal area; the circle has been *trisected.* How would you construct this figure using only a straightedge and compass? ______________________________

____

____

There are several ways to trisect a circle using a straightedge and compasses. In one of these methods, two circles are constructed inside and concentric with the given circle:

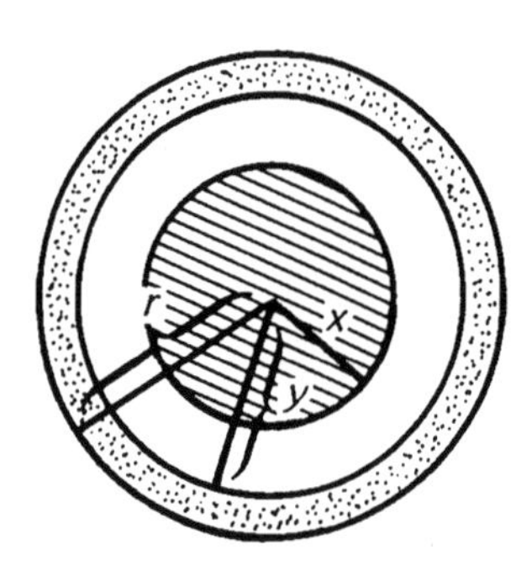

Each of the differently shaded regions is one-third the area of the largest circle. To trisect a circle this way, we must find x and y, the radii of the inner circles. First consider the circle with radius x. The area of this circle must be one-third the area of the given circle with radius r. Thus,

$$\pi x^2 = \frac{1}{3}\pi r^2,$$

$$x^2 = \frac{r^2}{3},$$

$$x = \frac{r}{\sqrt{3}} = \frac{r\sqrt{3}}{3}.$$

To construct x, write the last equation as $\frac{x}{\sqrt{3}} = \frac{r}{3}$. Using the unit segment and r, mark off lengths r and 3 on a line segment as shown (remember the two preceding units):

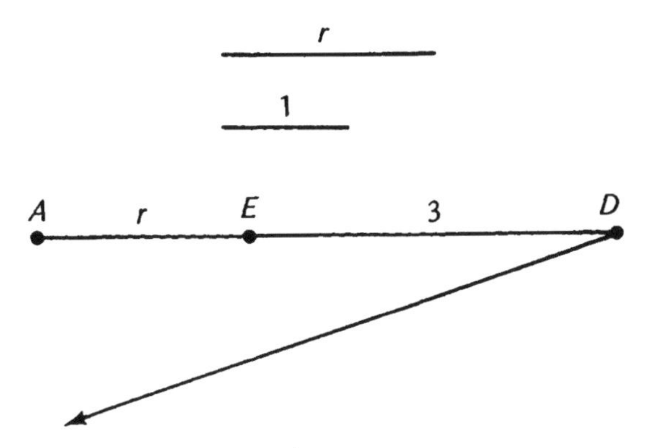

Then draw a ray from D. On a separate sheet of paper, use the given unit segment to construct a length $\sqrt{3}$. Mark off $DC = \sqrt{3}$ on the ray. Draw $\overline{EC}$. Construct a line through A parallel to $\overline{EC}$ and intersecting $\overrightarrow{EC}$ at B. What is the length of $\overline{BC}$? ____________. Why? ________________________

____

The same method is used to construct y, the radius of the other concentric circle. The area of the circle with radius y must be two-thirds the area of the circle with radius r. Why?________________________________

____

Therefore, $\pi y^2 = \frac{2}{3}\pi r^2$. What does y equal? ________________________
Construct y using the previously given r and unit segment. Then draw the three concentric circles using r and your constructed x and y.

EXTENSION! In the circle

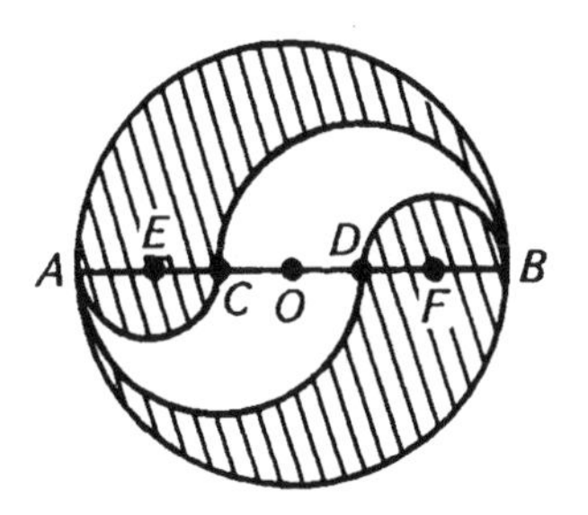

the diameter is trisected at C and D and four semicircles are drawn as shown. The circle is trisected into the two shaded regions and the unshaded region. Prove that this trisection is valid.

Teacher's Notes for Trisecting a Circle

In this activity students explore three methods for trisecting a circle. The method presented in the Extension is particularly interesting and will inspire some students to explore other circle constructions on their own. The activity reinforces basic construction techniques and gives students an opportunity to use the area formula for a circle in an interesting situation.

Students will need to use the constructions presented in "Constructing Segments" and "Constructing Radical Lengths," so these two activities should be reviewed if they have not been presented recently. Students should also know the area formula for a circle.

NCTM Standards

1	2	3	4	5	6	7	8	9	10
•	•				•	•		•	•

Presenting the Activity

The first trisection shown is easy to construct. Six equal arcs are marked off along the circle with the compass open to the radius of the circle. Radii are then drawn to every other mark.

Trisecting into concentric circles is more complicated and students must understand each step in the construction of x. First be sure they understand the figure: The areas of the small circle and each of the two rings must all be the same—each one equals one-third the area of the largest circle. Students must construct radius x and radius y given only r, the radius of the circle to be trisected. Using the formula for the area of a circle, students are shown how to find x in terms of r. Then the proportion

$$\frac{x}{\sqrt{3}} = \frac{r}{3}$$

is used to find x. The completed construction of x is

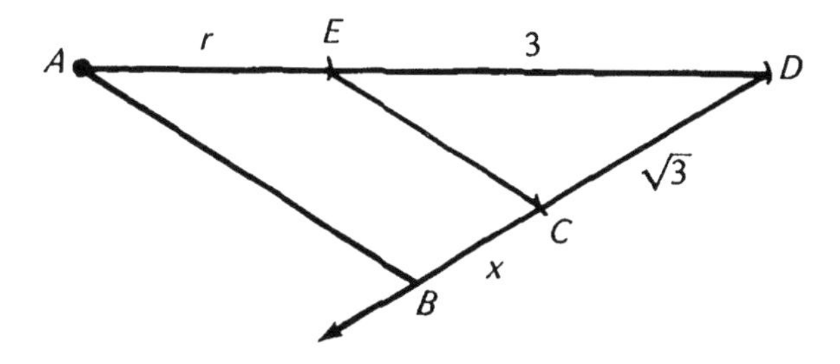

Students can use either of the methods from "Constructing Radical Lengths" to construct $\sqrt{3}$. The length of $\overline{BC}$ is x because a line parallel to one side of a triangle divides the other two sides proportionally.

Students may realize it's possible to construct

$$x = \frac{r\sqrt{3}}{3} \quad \left(\text{or} \quad x = \frac{r}{\sqrt{3}}\right)$$

using the methods shown in "Constructing Segments." You may wish to point out these other methods and ask students to demonstrate them.

Now students should turn their attention to the construction of y. The area of the circle with radius y must be two-thirds the area of the largest circle, so the area of the ring will be one-third the area of the largest circle. Some students may have difficulty seeing this. Remind them that the area of the unshaded ring and the area of the small circle must each be one-third the area of the largest circle. The area of the circle with radius y is the sum of the area of the unshaded ring and the area of the small circle. This sum is two-thirds the area of the largest circle:

$$\begin{aligned}\pi y^2 &= \frac{2}{3}\pi r^2,\\ y^2 &= \frac{2r^2}{3},\\ y &= \frac{r\sqrt{2}}{\sqrt{3}}\\ &= \frac{r\sqrt{6}}{3}.\end{aligned}$$

Students can then construct y as follows using the same method as was used to construct x:

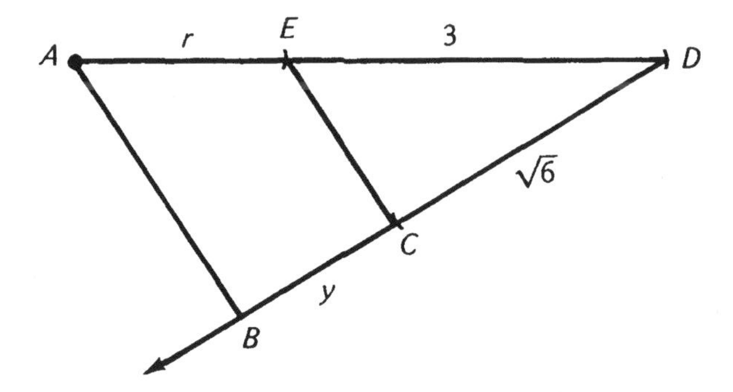

Extension

The trisection in the Extension is intriguing, but not difficult to prove valid; simply show that one of the shaded regions has an area one-third that of the original circle. The area of the upper shaded region = area semicircle AB − area semicircle BC + area semicircle AC. If $AE = r$, $AO = 3r$, and $BD = 2r$. Therefore, the area of the upper shaded region equals

$$\frac{1}{2}\pi(3r)^2 - \frac{1}{2}\pi(2r)^2 + \frac{1}{2}\pi r^2 = \frac{9}{2}\pi r^2 - \frac{4}{2}\pi r^2 + \frac{1}{2}\pi r^2 = 3\pi r^2.$$

The area of the original circle to be trisected equals $\pi(3r)^2 = 9\pi r^2$. Thus, the area of the shaded region is one-third the area of the original circle. You may wish to have students construct the figure on the student page as additional practice.

Trisecting an Angle

For hundreds of years mathematicians tried to trisect an angle using only a straightedge and a compasses. They were not successful. The reason for this lack of success was not explained until 1837. In that year algebra was used to prove that the construction is impossible. Of course, this doesn't mean it's impossible to trisect an angle using other methods.

One interesting method uses tools very close to a straightedge and compasses. The only difference is that the straightedge has two marks on it. These marks can be any convenient distance apart. The easiest way is to use a ruler and consider the marks to be, for example, 2 in. and 3 in. Now you can trisect $\angle AOB$:

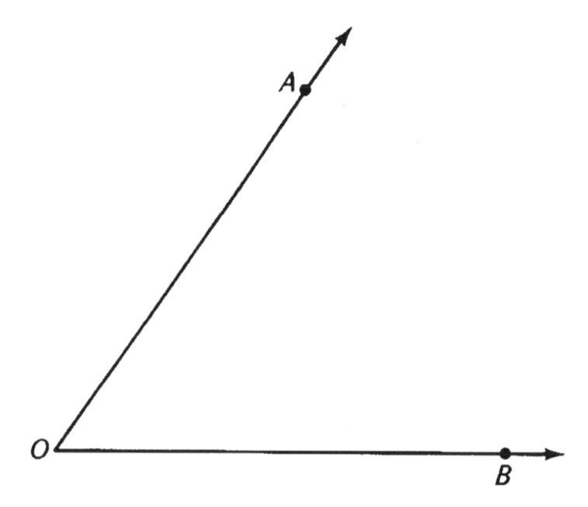

Open your compass to the distance between the two marks on your straightedge and draw a circle with the center at O. Label the point of intersection of the circle with $\overrightarrow{OA}$, C, and with $\overrightarrow{OB}$, D. Extend $\overline{DO}$ to intersect the circle at E and continue on farther. Now place your marked straightedge so that it contains C. At the same time have one mark on the circle and the other on $\overleftrightarrow{ED}$. Label the intersection with $\overleftrightarrow{ED}$ point F and draw line $\overleftrightarrow{FC}$. $m\angle CFD = \frac{1}{3}m\angle AOB$. Complete the construction by copying $\angle CFD$ twice in the interior of $\angle AOB$.

Now let's see if we can prove this construction is true. The figure

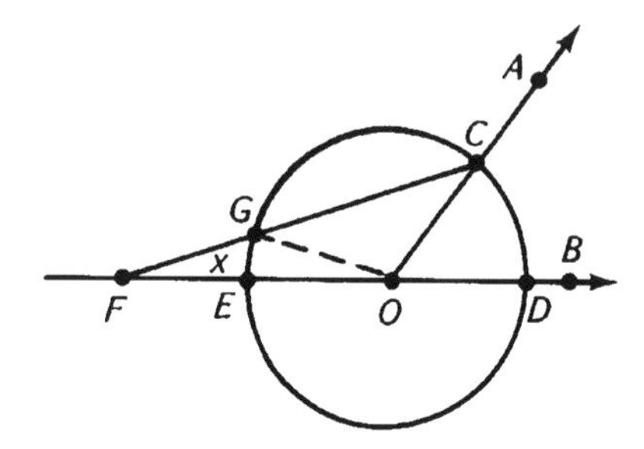

shows the results of your construction, with the addition of $\overline{GO}$.

Let $m\angle CFD = x$. We will have proved the trisection valid if $m\angle AOB = 3x$.

What is $m\angle GOF$?____________ Why? ______________________________

____

$\angle CGO$ is an exterior angle of $\triangle GOF$, so $m\angle CGO =$ ____________. Why?

____

What kind of triangle is $\triangle CGO$?____________ What is $m\angle GCO$? _______
$\angle AOB$ is an exterior angle of $\triangle CFO$, so $m\angle AOB =$ ____________.

Another method for trisecting an angle uses a tool called a tomahawk. A drawing of a tomahawk follows, but you should construct your own.

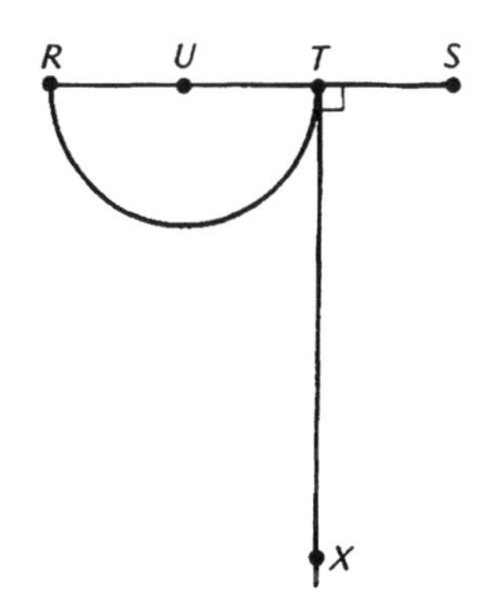

Start with a line segment $\overline{RS}$ trisected at U and T. Draw a semicircle about U with radius $\overline{UT}$ and construct $\overline{TX}$ perpendicular to $\overline{RS}$. To trisect any $\angle AOB$, simply place the tomahawk on the angle so that S falls on $\overrightarrow{OB}$, $\overline{TX}$ passes through vertex O, and the semicircle is tangent to $\overrightarrow{OA}$ at some point, say D. $\overrightarrow{OU}$ and $\overrightarrow{OT}$ trisect $\angle AOB$. Use tracing paper to trace $\angle AOB$ at the beginning of this activity. Then use this method to check that you get the same angle of trisection.

EXTENSION! Some angles *can* be trisected using only a straightedge and compasses. Explain how to trisect a 90° angle using straightedge and compasses. Then explain how this construction allows you to trisect a 45° angle.

Teacher's Notes for Trisecting an Angle

Probably the most widely known of the impossible geometric constructions attempted by the ancient Greeks is trisecting an angle. Unfortunately, students sometimes get the impression that the only way to trisect an angle is to use a protractor. However, this activity shows two interesting ways to trisect an angle using simple construction tools. In addition, the Extension shows that some specific angles ***can*** *be trisected using only a straightedge and a pair of compasses. In addition to giving construction practice, this activity provides practice in writing proofs.*

Students need to know basic constructions and should be familiar with the exterior angle theorem and basic properties of right triangles and circles.

NCTM Standards									
1	2	3	4	5	6	7	8	9	10
•	•				•	•		•	•

Presenting the Activity

Present the method most people think of when they try to trisect an angle using a straightedge and compasses. This gets it out of the way. In the accompanying figures, an arc from A is drawn intersecting the sides of the angle at B and C. Segment BC is drawn and then trisected at D and E. In the figure at the left, it appears $\overrightarrow{AD}$ and $\overrightarrow{AE}$ trisect $\angle BAC$, but this is not so. Try using an obtuse angle (figure at right, below) and the difference in the size of the angles will be obvious.

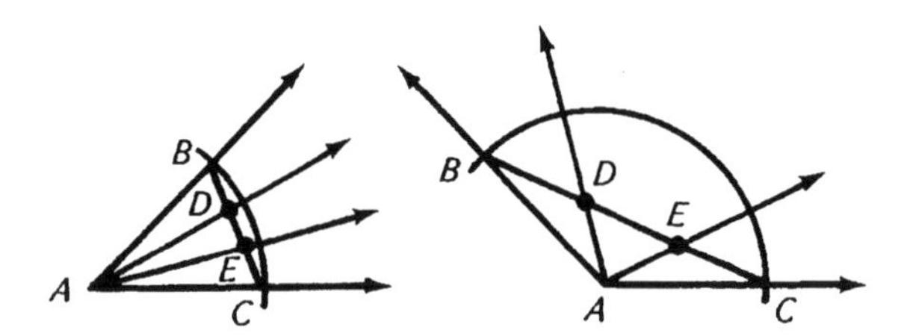

Now have students do the construction using compasses and a marked straightedge. To be sure they place the marked straightedge correctly, it may be necessary to do the construction on the chalkboard as students work at their desks. For additional practice, have students draw different angles and use this method to trisect them.

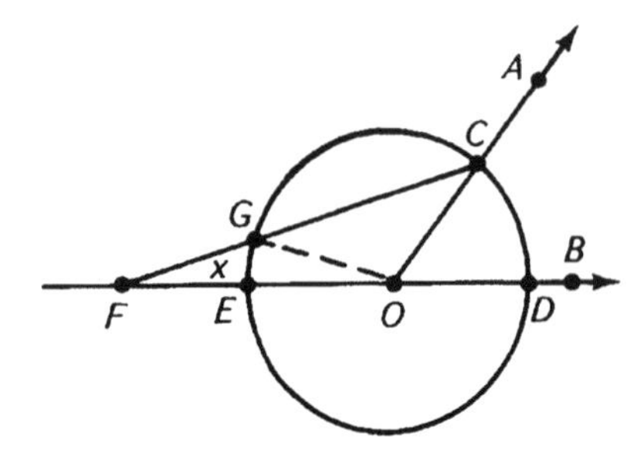

The proof following the construction is quite easy and students should not have trouble completing it. Whereas the radius of the circle is the same as the distance between the marks on the straightedge, $FG = OG$ and $\triangle GFO$ is isosceles. This means that $m\angle GOF = x$. The measure of an exterior angle of a triangle equals the sum of the measures of the two remote interior angles. Thus, $m\angle CGO = x + x = 2x$. Whereas $\triangle CGO$ is isosceles ($\overline{OG}$ and $\overline{OC}$ are radii), $m\angle GCO = m\angle CGO = 2x$, and because $\angle AOB$ is an exterior angle of $\triangle CFO$, so $m\angle AOB = m\angle GFO + m\angle FCO = x + 2x = 3x$. This shows that it doesn't matter what the measure of $\angle CFD$ is; the measure of $\angle AOB$ will be three times the measure of $\angle CFD$.

Emphasize the importance of a proof such as that given: If it can't be *proved* that a construction method is true, it can't be considered a valid method.

To use the tomahawk, the angle to be trisected must be drawn on tracing paper so it can be placed over the tomahawk. Thus, this is a good method to present the advantages of an overhead projector. The following figure below shows how $\angle AOB$ is placed over the tomahawk:

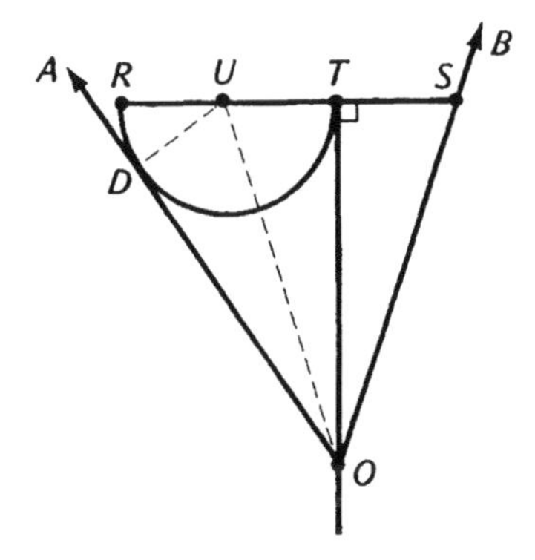

The proof that the tomahawk method is valid is also an easy one. Using the preceding figure, $\overline{UD}$ is drawn. Because they are radii of the same circle, $\overline{UD} \cong \overline{UT}$. Both $\angle UDO$ and $\angle UTO$ are right angles, so $\angle UDO \cong \angle UTO$. Because $\overline{UO} \cong \overline{UO}$, $\triangle UDO \cong \triangle UTO$ by the hypotenuse–leg congruence theorem. Also, $\overline{UT} \cong \overline{ST}$, $\angle UTO \cong \angle STO$, and $\overline{TO} \cong \overline{TO}$. Thus, $\triangle UTO \cong \triangle STO$ by the side–angle–side congruence theorem. Therefore, $\triangle UDO \cong \triangle UTO \cong \triangle STO$ and $m\angle DOU = m\angle TOU = m\angle TOS$.

Extension

To trisect a 90° angle using a straightedge and compasses, it's necessary to construct a 30° angle. Students can do this by constructing an equilateral triangle and bisecting one angle. Alternatively, they might use the converse of the 30-60-90 triangle theorem: If one leg of a right triangle is half as long as the hypotenuse, then the opposite angle has measure 30°. This construction is as follows: (1) Draw a base line and construct a perpendicular at some point C on the line. (2) Mark off a convenient length on the perpendicular and label this point A. (3) Using A as the center and $2AC$ as the radius, mark off an arc on the base line. Label this point B. (4) Draw $\overline{AB}$. Angle ABC is a 30° angle.

To trisect a 45° angle, it's necessary to construct a 15° angle. This is easily done by bisecting the 30° angle just constructed. Then add the 15° angle to the 30° angle to get a 45° angle.

You may wish to ask students to find other angles that can be trisected using a straightedge and a pair of compasses. Then have them explain how to do the construction.

Constructing a Pentagon

A regular pentagon has five sides of equal length. There is an easy way to make a regular pentagon using a strip of paper. Cut a strip of paper 1 in. wide and about 10 in. long. Tie the strip in a regular knot, pulling it taut and flattening it at the same time. Then cut off the excess flaps. If you hold the pentagon up to the light, you can see its diagonals forming a pentagram or star.

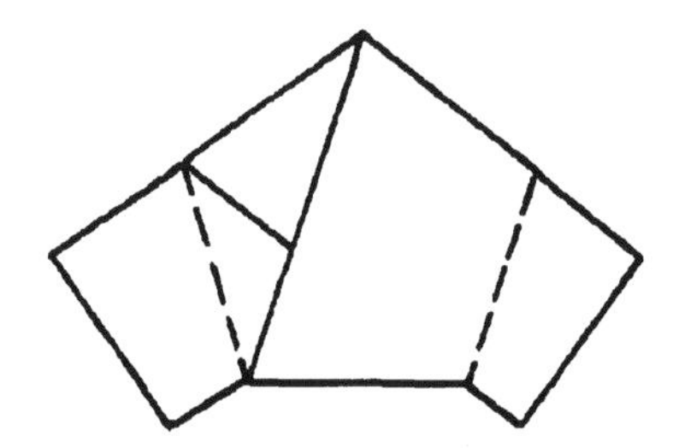

Constructing a regular pentagon with a straightedge and a pair of compasses is not as easy as knotting a strip of paper. In the following figure, a point A was selected on line ℓ, a perpendicular was constructed at A, and a point O was selected on the perpendicular:

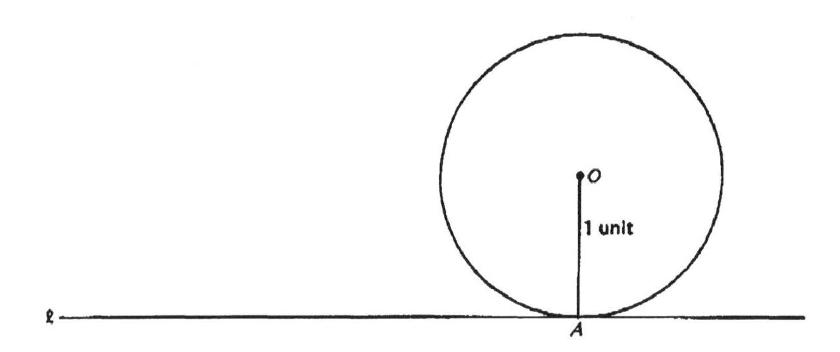

The length of $\overline{OA}$ is 1 unit. Then the circle with center O and radius OA was drawn. Using the figure, complete the construction as follows: Mark point P on ℓ such that AP is 2 units. Draw $\overline{OP}$ and label the intersection of $\overline{OP}$ and the circle point Q. Construct the midpoint R of $\overline{PQ}$. PR is the length of a side of the regular decagon (10-sided polygon) inscribed in the circle.

How can you use $\overline{PR}$ to construct the pentagon? ____________________

____

____

____

Complete the construction.

EXTENSION! The following figure

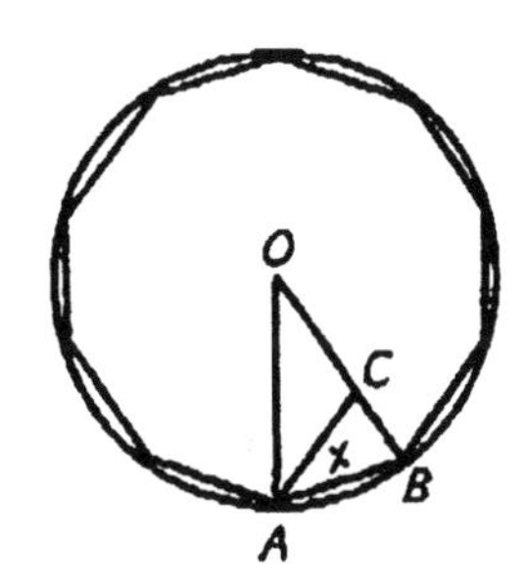

is an inscribed regular decagon. $\overline{OA}$, $\overline{OB}$, and $\overline{AB}$ form an isosceles triangle and $\overline{AC}$ bisects $\angle OAB$. Let $AB = x$ and use similar triangles to show that $x = \frac{\sqrt{5}-1}{2}$. Then use the Pythagorean theorem to show that $\overline{PR}$ in the figure you previously constructed has length $\frac{\sqrt{5}-1}{2}$.

Teacher's Notes for Constructing a Pentagon

This activity has a twofold purpose. First, it presents a method for constructing a pentagon and reinforces simple construction techniques. Second, the Extension explores why the construction works and uses algebraic formulas to discover geometric relationships. The construction method is different from the method usually presented in geometry texts and is much easier to justify—an important consideration for any construction.

Students should be familiar with the properties of inscribed regular polygons and similar triangles. The activity is best presented either immediately before or after "The Golden Rectangle" and "The Golden Triangle."

NCTM Standards

1	2	3	4	5	6	7	8	9	10
•	•	•	•		•	•		•	•

Presenting the Activity

It's a good idea to have strips of paper cut before class begins. Tying the strip of paper in a knot and flattening it is quite easy and most students will be surprised that they haven't seen it before.

Students should have no difficulty with the straightedge and compass construction. Be sure they understand that the figure shown on the student student page was constructed using a straightedge and a compass. It is given just to help them get started. If necessary, remind students that the midpoint R is found by constructing the perpendicular bisector of $\overline{PQ}$.

By marking 10 segments of length PR on the circle, students will determine the vertices of the inscribed decagon. Then they can join alternate vertices of the decagon to draw the pentagon. The complete construction is

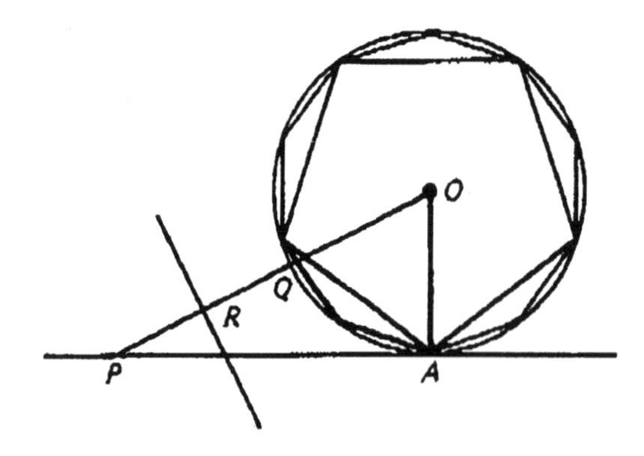

Extension

Although many of the Extensions in this volume are not essential to the development of a topic, this one is an exception. Whereas the construction of a regular pentagon is useful by itself, the justification of why it works helps develop a better understanding of geometric concepts and their relationship to algebra.

Many students will find the Extension difficult to do on their own, so be sure to allow enough time to present it as a class discussion. Begin by drawing isosceles triangle AOB on the chalkboard. Students should easily see that $m\angle AOB = 36°$, because $\frac{360°}{10} = 36°$. Therefore, $m\angle OAB = m\angle OBA = 72°$. Whereas $\overline{AC}$ bisects $\angle OAB$, $m\angle OAC = m\angle CAB = 36°$. Thus, $\triangle OCA$ and $\triangle CAB$ are isosceles and we have the figure

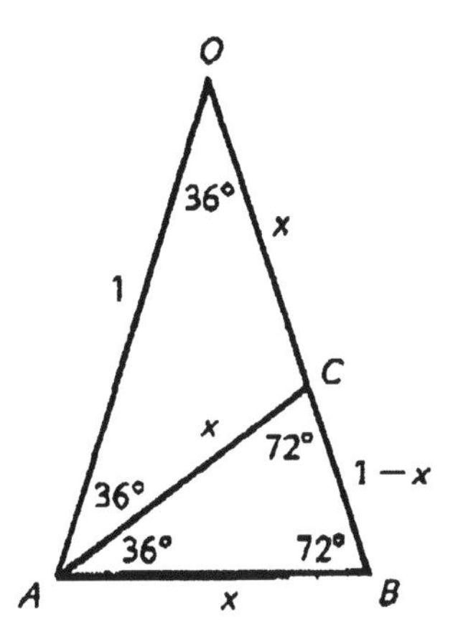

Because $\triangle AOB \sim \triangle BAC$, we can write the proportion

$$\frac{1}{x} = \frac{x}{1-x},$$

which leads to the equation $x^2 + x - 1 = 0$. Solving for x using the quadratic formula, we find one root is

$$x = \frac{\sqrt{5}-1}{2}.$$

This completes the first half of the justification.

Now it's necessary to show that $\overline{PR}$ in the construction has length x. Because $OA = 1$ and $AP = 2$, by the Pythagorean theorem, $OP = \sqrt{5}$. Thus, $PQ = OP - OQ = \sqrt{5} - 1$. Because R is the midpoint of $\overline{PQ}$,

$$PR = \frac{\sqrt{5}-1}{2} = x.$$

This justifies the construction because a *converse* argument can be made: Whereas

$$PR = AB = \frac{\sqrt{5}-1}{2},$$

$m\angle AOB = 36°$ and is therefore one central angle of a regular decagon.

If students have studied "The Golden Rectangle" and "The Golden Triangle," they should recognize x as ϕ', the reciprocal of the golden ratio, and $\triangle AOB$ as a golden triangle. If they have not studied these activities, this Extension is an excellent introduction to them.

Constructing Triangles

You have used the ancient Greek geometers' straightedge and compasses to construct triangles given the measures of various sides and angles. Triangles can also be constructed given the measures of other parts of triangles, such as altitudes, medians, and angle bisectors. Let's consider constructing a triangle given the lengths of two sides and the altitude to one of these sides. These lengths are given on page 28 (h_a is the altitude to side a).

It's usually a good idea to sketch the "finished product" before beginning the construction. In many cases this suggests an idea of where to start. A sketch of the finished triangle with the given lengths in bold lines is

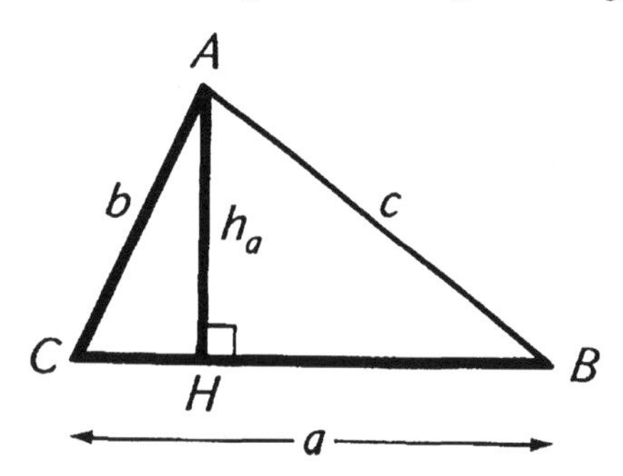

Use the base line on page 28 to do the following constructions:

1. Construct a perpendicular at H.
2. Mark off the segment $\overline{HA}$ of length h_a on this perpendicular.
3. With A as the center and b as the radius, draw the arc that intersects the base line. Label this point C.
4. With C as the center and a as the radius, draw the arc that intersects the base line. Label this point B.
5. Draw $\triangle ABC$.

Is $\triangle ABC$ unique? That is, is there only one triangle with sides a and b and altitude h_a?________________________________

Under what conditions will there be only one triangle?________________

Under what conditions will there be no triangle?____________________

Now let's construct a triangle given only the lengths of its medians. This is more difficult and it's best to begin by analyzing the construction in reverse. Again we start with a sketch of the finished triangle.

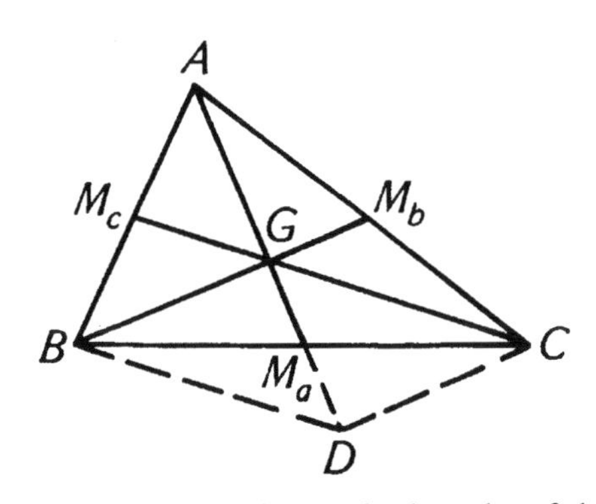

Because we don't know the lengths of the sides of $\triangle ABC$ or the measures of any angles, we can't construct any of the triangles in the figure. So we extend $\overline{GM_a}$ its own length to D and draw $\overline{BD}$ and $\overline{CD}$ as shown. What kind of quadrilateral is $BGCD$?____________________

Why?____________________

How can you determine the lengths of the sides of $\triangle GBD$?____________________

____

Use the given lengths of medians m_a, m_b, and m_c on page 28 and construct $\triangle GBD$.

How can you locate point M_a?____________________

How can you locate point C?____________________

How can you locate point A?____________________

____

Complete the construction.

EXTENSION! Construct $\triangle ABC$ given the lengths of a, the altitude to b, and the median to c as shown:

a

h_b

m_c

Teacher's Notes for Constructing Triangles

Although most geometry students are familiar with the basic triangle constructions, they have usually not been exposed to any of the more difficult ones. This activity presents three interesting triangle constructions that not only review geometric concepts, but also help students analyze geometry problems.

Students should be familiar with the basic geometry constructions and the median concurrency theorem. The segments to be used in the constructions are given in reproducible form on page 28.

NCTM Standards

1	2	3	4	5	6	7	8	9	10
•	•	•	•		•	•	•	•	•

Presenting the Activity

Begin by discussing the triangle constructions students have already done in terms of the given angles and sides. Students should have constructed triangles given the measures of

1. two sides and their included angle
2. two sides and any angle
3. two angles and their included side
4. two angles and any side
5. three angles
6. three sides.

Ask students which of these constructions produce unique triangles and which do not (2 and 5 are not unique). Be sure they understand that for a triangle construction to *not* produce a unique triangle, the two triangles produced must *not* be congruent.

Now discuss the sketch for the first construction. Ask students for ideas on how to do the construction. They should see that $\triangle ACH$ can be constructed easily because they have the lengths of the hypotenuse and leg of a right triangle. (Point out that $\triangle ACH$ is a special case of construction 2 above, and *does* produce a unique triangle.) Then have students follow the step-by-step constructions to draw $\triangle ABC$.

There should be one or two students who realize $\triangle ABC$ is not unique. If not, refer students to construction step 4 on the student page, pointing out that the arc can cross the base line on either side of point C. The completed construction is shown on page 29.

Have students consider the lengths of b and h_a to answer the next two questions on the student page. If $b = h_a$, $\triangle ABC$ will be unique and if $b < h_a$, the construction is not possible.

The next construction is much more difficult. By extending $\overline{GM_a}$ its own length to D, we have $GM_a = DM_a$. Also, M_a is the midpoint of $\overline{BC}$, so $BM_a = CM_a$. Therefore, $BGCD$ is a parallelogram because the diagonals bisect each other. To determine the lengths of the sides of $\triangle GBD$, students need to know that the medians of a triangle meet in a point that is the trisection of each one. Thus, $BG = \frac{2}{3}m_b$, $BD = GC = \frac{2}{3}m_c$, and, because $GM_a = DM_a = \frac{1}{3}m_a$, $GD = \frac{2}{3}m_a$.

After constructing $\triangle GBD$, students can find M_a by bisecting $\overline{GD}$. They can locate C by extending $\overline{BM_a}$ and marking off $\overline{M_aC}$ equal in length to $\overline{BM_a}$. Point A can be located several different ways. Extending $\overline{GD}$ and marking off $\overline{M_aA}$ equal in length to m_a is one way. Alternatively, $\overline{BG}$ can be extended and $\overline{BM_b}$ equal in length to m_b marked off, $\overline{CG}$ can be extended and $\overline{CM_c}$ equal in length to m_c marked off, and $\overline{BM_c}$ and $\overline{CM_b}$ extended to meet at A. The completed construction is shown on page 29.

Not only does this problem review many important concepts from elementary geometry, but it also provides an opportunity for students to practice "reverse" reasoning to analyze the problem.

Extension

Students should again make a sketch of the triangle as shown:

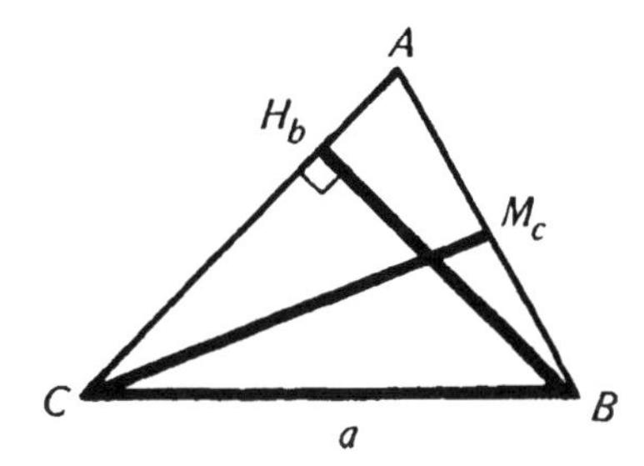

They can construct $\triangle CBH_b$ using the same method described in the first construction on the student page: Draw any line AC, choose a point H_b, and construct a perpendicular at H_b of length h_b. Then from B with radius a locate point C on line AC. Drawing $\overline{BC}$ completes the construction of $\triangle CBH_b$.

Before continuing, students should again consider their original sketch. If they can construct a parallelogram $ADBC$ as shown,

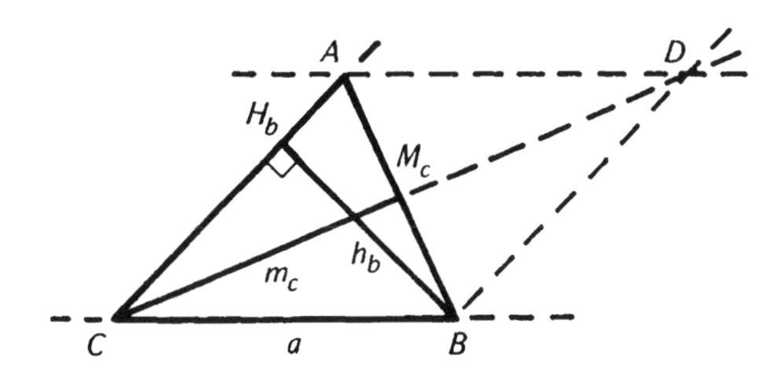

they will be able to complete the construction of $\triangle ABC$.

This parallelogram is constructed as follows: (1) A line through B is constructed parallel to line AC. (2) Using C as the center and $2m_c$ as the radius, point D is marked on the line through B. (3) Point A can be found by constructing a line through D parallel to $\overline{CB}$, or by marking off length m_c on $\overline{CD}$ and extending $\overline{BM_c}$ to line AC.

When you consider the measures of other parts of a triangle, such as angle bisectors, radius of an inscribed circle, radius of a circumscribed circle, and the semiperimeter as well as altitudes, medians, angles, and sides, then there are 179 possible triangle construction problems. Each consists of measures of three of these parts of a triangle.

a

b

h_a

H

m_a

m_b

m_c

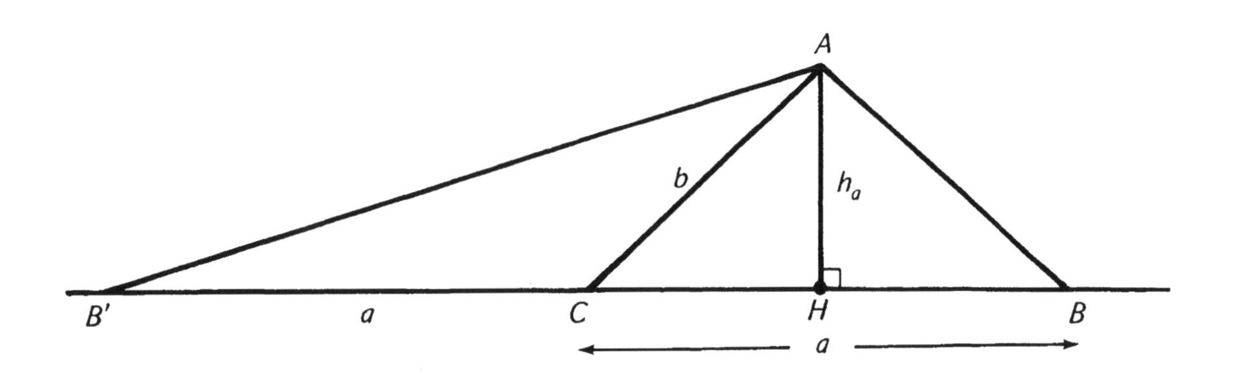
A
b
h_a
B'
a
C
H
B
a

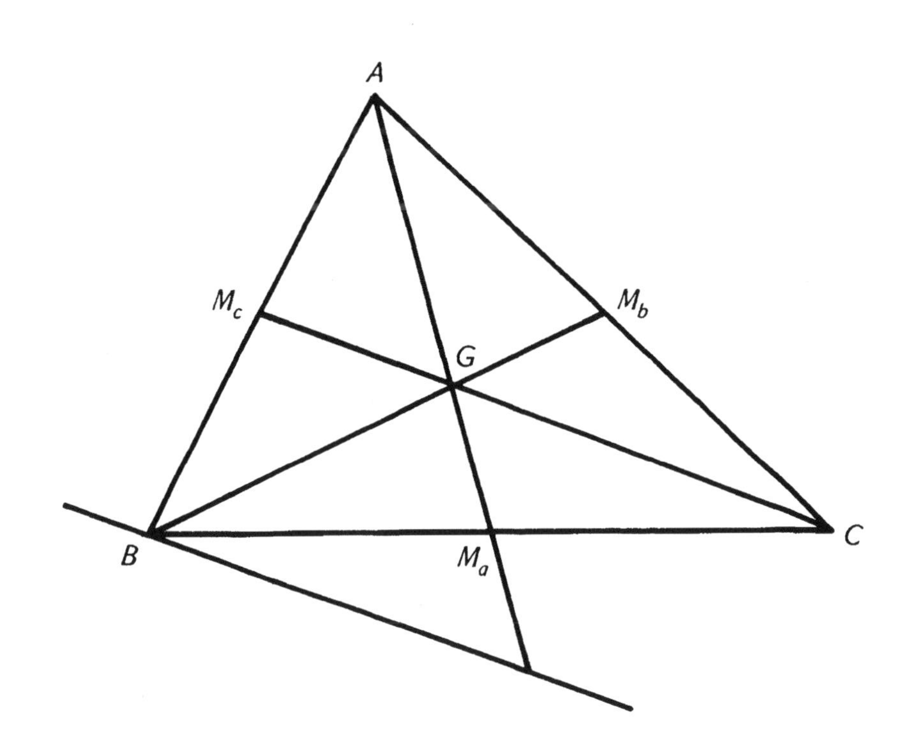
A
M_c
M_b
G
B
M_a
C

CHAPTER 2

Problems of Antiquity

- The Pythagorean Theorem
- The Golden Rectangle
- The Golden Triangle
- The Arbelos

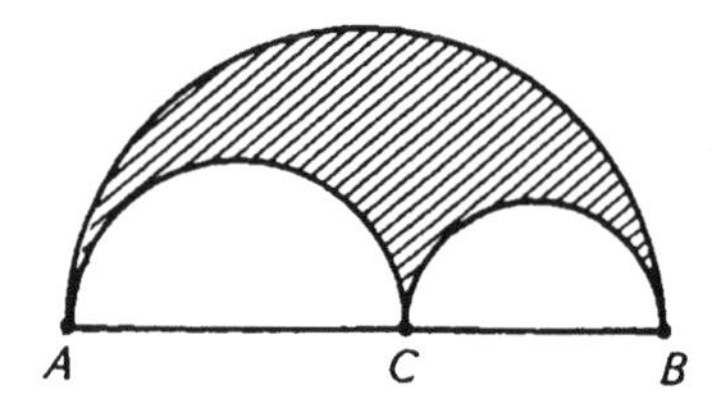

The ancient Greeks considered many geometric problems. These problems and the relationships developed in their solutions are the basis of Euclid's *Elements*. Because geometry is taught as a logical mathematical system—one relationship proved from an earlier proven relationship—most students don't realize that geometry did not somehow spring into being as a finished, logical system. It grew from considering specific problems in art, architecture, and nature. Thus, consideration of some of the problems of antiquity can help students realize why the Greeks found geometry fascinating and beautiful. However, many of the problems of antiquity are not included in geometry texts. The four activities in this category consider three very different geometric problems and develop many geometric and algebraic relationships.

The first activity considers the most famous relationship discovered by the Greeks, the Pythagorean theorem. This theorem is proved in every geometry text, usually using similar triangles, as shown in the first method on the student page. This proof was discovered by Pythagoras. The other two proofs in "The Pythagorean Theorem" are also historically interesting because of the people who developed them—Euclid and President James A. Garfield. These two proofs are based on area. The three proofs given here emphasize that there is often more than one correct proof of a geometric theorem and more than one way to solve a geometric problem.

"The Golden Rectangle" can be a good change of pace early in a geometry course. The Greeks were unaware of the algebraic relationships in a golden rectangle and considered it only for its pleasing proportions. By incorporating the algebraic properties discovered about 2000 years later, "The Golden Rectangle" presents an interesting example of the relationship between algebra and geometry.

"The Golden Triangle" extends some of the properties of a golden rectangle to a triangle. Finding the measures of the angles of a golden triangle requires some knowledge of circles, so this activity is best presented later in the year. The link between a golden triangle and a golden rectangle is provided by the golden ratio and the golden spiral.

Most geometry texts present triangles, quadrilaterals, and circles as separate topics. The important theorems and definitions are developed for each type of figure separately. As a result, few problems are given whose solutions depend on properties from all three types of figures. The relationships developed in "The Arbelos" use properties of triangles, rectangles, and circles, thus applying students' knowledge in several areas. "The Arbelos" shows students how several relationships can be developed from a relatively simple geometric figure. This, of course, was a large part of the Greeks' fascination with geometry.

The Pythagorean Theorem

More different proofs have been written for the Pythagorean theorem than for any other theorem in geometry. One book lists 370 of them. Pythagoras is given credit for developing the following proof based on similar triangles.

Given right $\triangle ABC$ with the right angle at C, prove $a^2 + b^2 = c^2$.

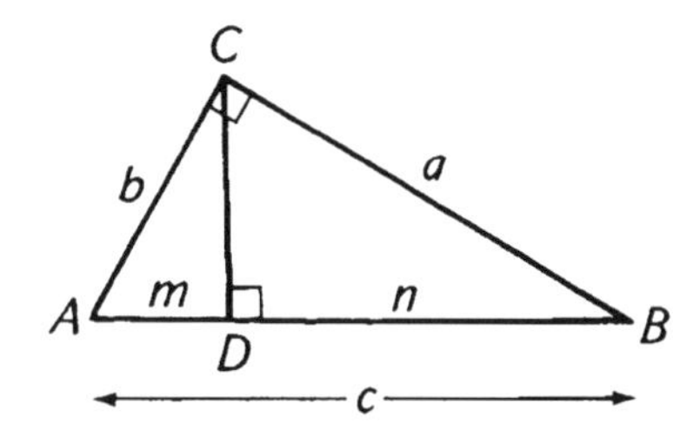

Altitude $\overline{CD}$ forms two right triangles that are similar to each other and to the original triangle. Therefore,

$$\frac{c}{b} = \frac{b}{m} \quad \text{or} \quad \frac{c}{a} = \frac{a}{n}.$$

So, $b^2 =$ ________ and $a^2 =$ ________. Adding, $a^2 + b^2 =$ ____________.

Complete the proof: ____

Another proof of the Pythagorean theorem was discovered by President James A. Garfield in 1876 while he was a member of Congress. Garfield's proof is based on area.

First trapezoid $DEBC$ is constructed using the lengths from $\triangle ABC$, as shown:

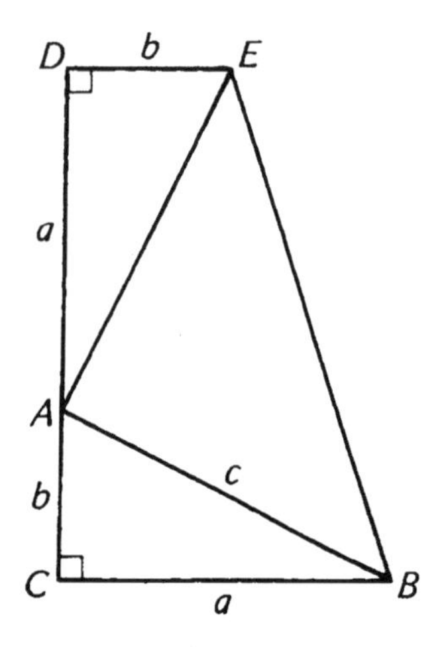

What is the relationship between $\triangle EAD$ and $\triangle ABC$? ____________________

What is the length of $\overline{AE}$? _______ What kind of angle is $\angle EAB$? _______
Why? ____

____

Now consider the areas of the trapezoid and the three triangles. What is the formula for the area of a trapezoid? ______________________________
What is the formula for the area of a triangle? ______________________
Therefore,

$$\text{area}\, DEBC = \text{area}\, \triangle EAD + \text{area}\, \triangle EAB + \text{area}\, \triangle ABC.$$

Find the preceding areas in terms of a, b, and c:

area $DEBC =$ ____________; area $\triangle EAD =$ ____________;

area $\triangle EAB =$ ____________; area $\triangle ABC =$ ____________.

Use these areas and the relationship among them to complete the proof.______

____

____

EXTENSION! Another proof of the Pythagorean theorem based on area was developed by Euclid. In the accompanying figure, squares are constructed on the sides of right $\triangle ABC$. $\overline{LC}$ is constructed perpendicular to $\overline{DE}$, and $\overline{CE}$ and $\overline{KB}$ are drawn. Prove that area $ACHK$ + area $BCGF$ = area $ABDE$.

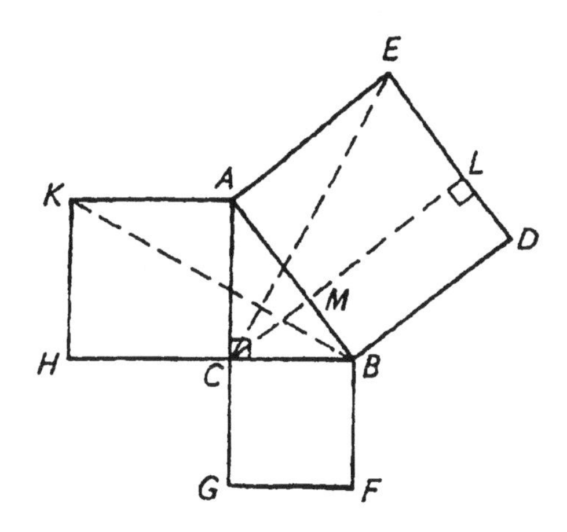

Teacher's Notes for The Pythagorean Theorem

Although all geometry texts prove the Pythagorean theorem, most students are unaware of the variety of existing proofs. Three proofs are presented in this activity—two using area relationships and one using proportions in similar triangles. Thus, a wide range of geometric concepts are reinforced. In addition, the proofs are historically interesting: One is by Pythagoras, the originator of the theorem; one is by Euclid, one of the most famous mathematicians of all times; one is, unexpectedly, by President James A. Garfield.

The only prerequisites for the activity are knowledge of the area formulas and proportional segments in right triangles.

NCTM Standards

1	2	3	4	5	6	7	8	9	10
•	•	•	•		•			•	•

Presenting the Activity

Students will be amazed that more than 370 different proofs exist for one theorem. The book referred to is *The Pythagorean Proposition* by Elisha S. Loomis (National Council of Teachers of Mathematics, Washington, DC, 1968). Other proofs are continuously published in journals, since the first edition of this book in 1940.

Given the proportions to use in Pythagoras' proof, students should be able to complete the proof easily:

$$b^2 = cm \quad \text{and} \quad a^2 = cn,$$

$$\begin{aligned} a^2 + b^2 &= cn + cm \\ &= c(n+m) \\ &= c \cdot c, \\ a^2 + b^2 &= c^2. \end{aligned}$$

In Garfield's proof, $\triangle EAD \cong \triangle ABC$ and the length of $\overline{AE}$ is c. Because $\triangle EAD \cong \triangle ABC$, $\angle EAD \cong \angle ABC$. In $\triangle ABC$, $m\angle ABC + m\angle CAB = 90°$. Therefore, $m\angle EAD + m\angle CAB = 90°$ and $\angle EAB$ is a right angle.

The area of a trapezoid is $\frac{1}{2}h(m+n)$, where h is the altitude and m and n are the lengths of the bases. The area of a triangle is $\frac{1}{2}hn$, where h is the altitude and n is the length of the base. (Note that these area formulas are usually given using b to represent the length of the base. It may be best for students to use letters other than b to avoid confusion with length b in the figure.) The areas in terms of a, b, and c are:

$$\begin{aligned} \text{area } DEBC &= \frac{1}{2}(a+b)(a+b), \\ \text{area } \triangle EAD &= \frac{1}{2}ab, \\ \text{area } \triangle EAB &= \frac{1}{2}cc, \\ \text{area } \triangle ABC &= \frac{1}{2ab}. \end{aligned}$$

Therefore,

$$\begin{aligned}
\frac{1}{2}(a+b)(a+b) &= \frac{1}{2}ab + \frac{1}{2}c^2 + \frac{1}{2}ab, \\
\frac{1}{2}(a^2+2ab+b^2) &= \frac{1}{2}(c^2+2ab), \\
a^2+b^2+2ab &= c^2+2ab, \\
a^2+b^2 &= c^2.
\end{aligned}$$

Extension

Many students will find Euclid's proof difficult. If it is not worked through in class, it should only be assigned to better students. Point out that the statement to be proved is the geometric equivalent of $a^2+b^2=c^2$.

Students should begin by proving $\triangle CAE \cong \triangle KAB$ ($\overline{CA} \cong \overline{KA}$, $\overline{AE} \cong \overline{AB}$, and $\angle CAE \cong \angle KAB$). By redrawing only parts of the figure on the student page as the following diagrams show, it is easier to see the relationships:

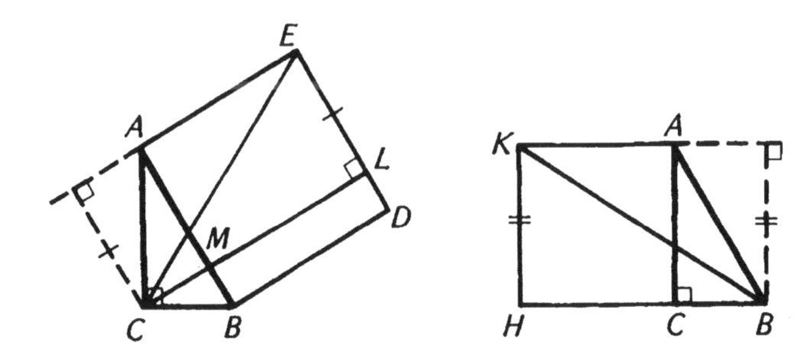

In the figure on the left, $\triangle CAE$ and rectangle $LEAM$ have the same base, $\overline{AE}$. The altitude to $\overline{AE}$ in $\triangle CAE$ (shown by dashed lines) is congruent to $\overline{EL}$ and $\overline{AM}$, the altitudes of rectangle $LEAM$. Thus,

$$\text{area}\,\triangle CAE = \frac{1}{2}\,\text{area}\,LEAM.$$

Similarly, using the right-hand figure, $\triangle KAB$ and square $ACHK$ have the same base and congruent altitudes, so

$$\text{area}\,\triangle KAB = \frac{1}{2}\,\text{area}\,ACHK.$$

Because $\triangle CAE \cong \triangle KAB$,

$$\begin{aligned}
\text{area}\,\triangle CAE &= \text{area}\,\triangle KAB, \\
\text{area}\,LEAM &= \text{area}\,ACHK.
\end{aligned}$$

Now students must draw $\overline{CD}$ and $\overline{AF}$:

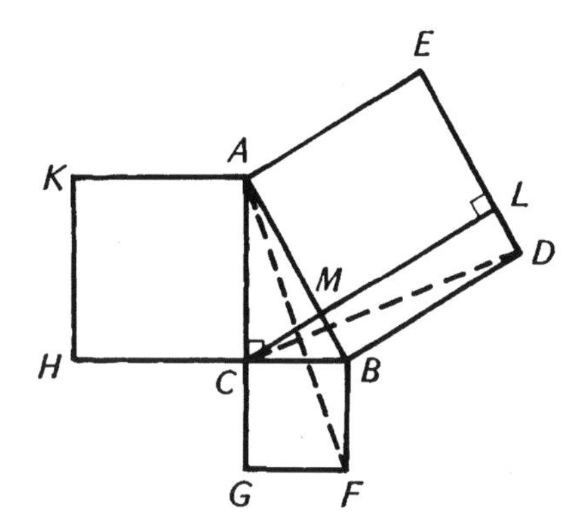

Then, using $\triangle ABF$ and $\triangle DBC$, they can prove

$$\text{area } BCGF = \text{area } BDLM$$

using the same method as before. By addition,

$$\text{area } ACHK + \text{area } BCGF = \text{area } LEAM + \text{area } BDLM$$

and

$$\text{area } ACHK + \text{area } BCGF = \text{area } ABDE.$$

The Golden Rectangle

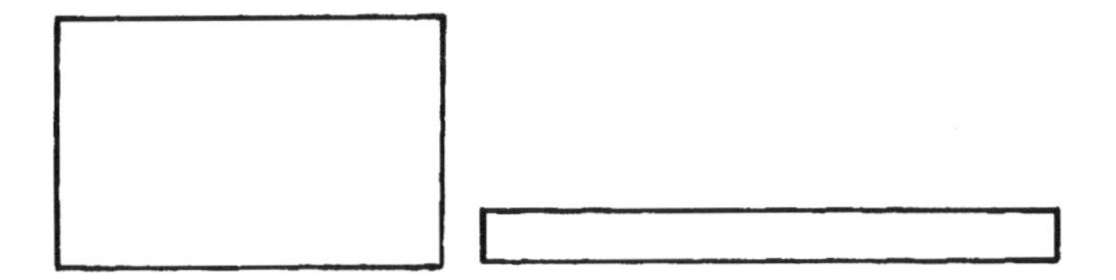

Which of the preceding rectangles is more pleasing to look at? According to the ancient Greeks, and verified by modern psychologists, most people find the rectangle on the left more pleasing. The Greeks used this rectangle in constructing many of their buildings—it is called a *golden rectangle*.

Follow these steps to construct a golden rectangle such as $ABFE$:

1. Construct a square about 10 cm on a side and label it $ABCD$.
2. Locate the midpoint M of $\overline{AD}$ and draw $\overline{MC}$.
3. With a compass centered at M, and having radius MC, draw an arc from C to meet $\overrightarrow{AD}$ at E.
4. At E, construct a perpendicular to $\overline{AE}$ that meets $\overrightarrow{BC}$ at F.

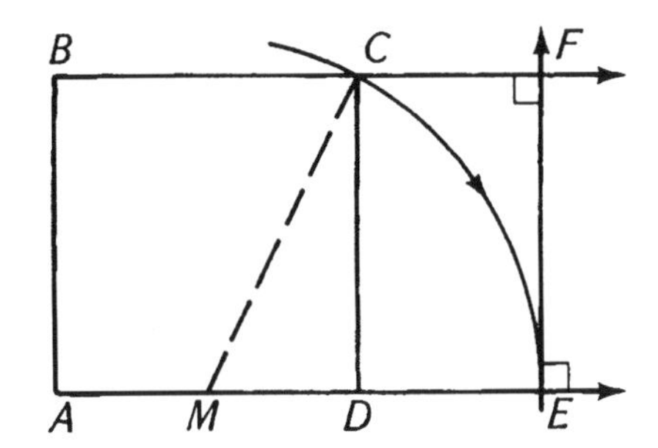

The ratio of the length, AE, of a golden rectangle to its width, EF, is called the *golden ratio*. Let's find this ratio. Let $AD = 1$ unit. Then $MD =$ _______. Using $\triangle MDC$ and the Pythagorean theorem, $MC =$ _______.

Whereas we constructed $ME = MC$, $ME =$ _______. Thus, $AE = AM + ME =$ ___________, and, whereas $EF = AD = 1$, $\frac{AE}{EF} =$ ___________.

The golden ratio is denoted by the Greek letter ϕ (phi). Thus, $\phi = \frac{\sqrt{5}+1}{2}$. Now look at rectangle $CDEF$. Is it a golden rectangle? We can find out by finding the ratio of length (EF) to width (DE):

$EF =$ ___________ and $DE = ME - MD =$ _______.

Thus, $\frac{EF}{DE} =$ _______. Is this equal to ϕ? _______

Is $CDEF$ a golden rectangle?______________________________

Now draw $\overline{HG}$ as shown in the following figure so that $DEGH$ is a square. $CFGH$ is also a golden rectangle.

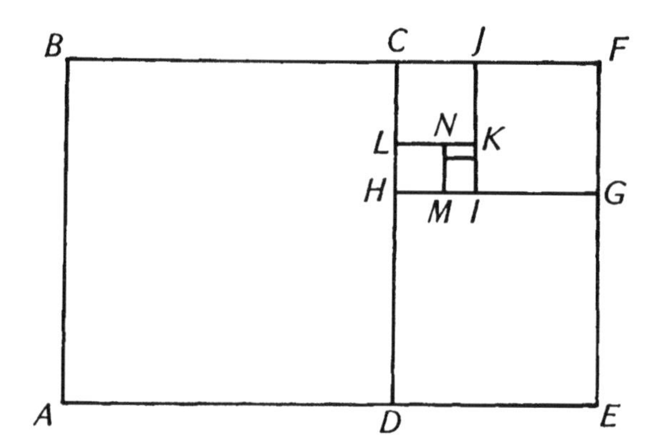

Continue partitioning the smaller rectangle into a square and another golden rectangle as shown in the figure. Then, using a compass with C as the center and BC as radius, draw arc BD. Use H as the center and HD as radius, and draw arc DG. Use I as the center and draw arc GJ. Continue drawing arcs to form a spiral. This spiral is closely related to the spiral of the chambered nautilus shell and is called the *golden spiral*.

There are many interesting mathematical relationships that involve ϕ. To discover one of them, find ϕ', the ratio of *width to length* in a golden rectangle. Then find the product and difference of ϕ and ϕ'.

$$\phi' = \frac{1}{\phi} = \_\_\_\_\_\_, \qquad \phi \cdot \phi' = \_\_\_\_\_\_, \qquad \phi - \phi' = \_\_\_\_\_\_.$$

Are there any other numbers in mathematics for which this is true? If so, what are they? ______________________________

EXTENSION! Another interesting mathematical relationship concerns the powers of ϕ. First, show that $\phi^2 = \phi + 1$. Then use this equation to write the powers of ϕ from ϕ^1 to ϕ^8 in terms of ϕ. What do you notice about the coefficients and the constants? What is this sequence called?

Teacher's Notes for The Golden Rectangle

Very few topics in algebra and geometry present as many fascinating relationships as the golden rectangle. Discovery of these relationships spanned many centuries, for although the ancient Greeks knew the geometric proportions of the golden rectangle, its algebraic relationships could not have been known until almost 2000 years later. This activity reinforces construction techniques and relates geometry to algebra. The only geometric prerequisites are constructing midpoints and perpendiculars.

*The Extension for this activity relates geometry to algebra via the Fibonacci sequence. For a more detailed consideration of this topic, see **Making Algebra Come Alive**.*

NCTM Standards

1	2	3	4	5	6	7	8	9	10
•	•	•	•		•	•		•	•

Presenting the Activity

Briefly discuss the two rectangles at the top of the first student page. Point out that the rectangle at the right requires a scanning motion in viewing, whereas the one at the left can be appreciated at one glance. If pictures of ancient Greek buildings (e.g., the Parthenon) are available, compare the golden rectangle to the shape of these buildings. Then have students do the construction indicated. (*Note*: Square $ABCD$ can be any size—10 cm on each side is suggested because it is a convenient measure.)

Students should not have any difficulty finding the golden ratio. Because M is the midpoint of $\overline{AD}$, $AM = MD = \frac{1}{2}$. Also, $ME = MC = \frac{\sqrt{5}}{2}$, so

$$AE = AM + ME = \frac{1}{2} + \frac{\sqrt{5}}{2} = \frac{\sqrt{5}+1}{2}.$$

Therefore, $\frac{AE}{EF} = \frac{\sqrt{5}+1}{2}$.

Rectangle $CDEF$ is also a golden rectangle because the ratio of length to width is ϕ as shown by

$$\begin{aligned} DE &= ME - MD \\ &= \frac{\sqrt{5}}{2} - \frac{1}{2} = \frac{\sqrt{5}-1}{2}, \\ \frac{EF}{DE} &= \frac{1}{\frac{\sqrt{5}-1}{2}} = \frac{2}{\sqrt{5}-1}. \end{aligned}$$

Rationalizing the denominator gives

$$\frac{EF}{DE} = \frac{\sqrt{5}+1}{2} = \phi.$$

When a square is cut from one end of a golden rectangle, the remaining rectangle is also a golden rectangle. This idea is explored in the next construction. Successively smaller golden rectangles are formed by partitioning squares from golden rectangles. If time permits, students can verify that these smaller rectangles are golden by finding the ratio of length to width. The completed figure with the spiral drawn is

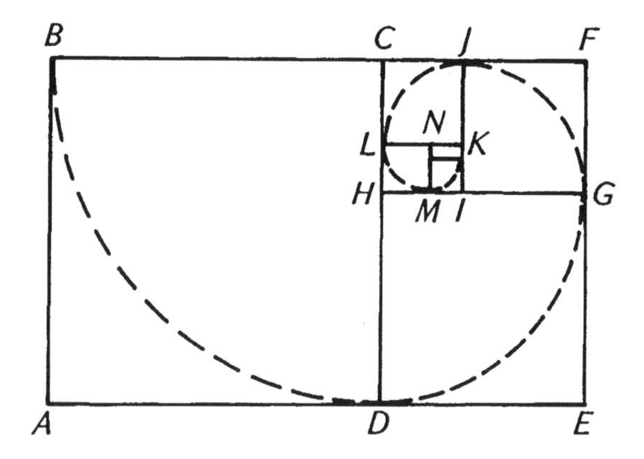

The relationship between ϕ and its reciprocal ϕ' is developed next:

$$\phi' = \frac{1}{\phi} = \frac{1}{\frac{\sqrt{5}+1}{2}} = \frac{\sqrt{5}-1}{2},$$

$$\phi \cdot \phi' = \frac{\sqrt{5}+1}{2} \cdot \frac{\sqrt{5}-1}{2} = \frac{5-1}{4} = 1,$$

$$\phi - \phi' = \frac{\sqrt{5}+1}{2} - \frac{\sqrt{5}-1}{2} = 1.$$

To check for other numbers, students should consider two numbers, x and y, such that $xy = 1$ and $x - y = 1$. Combining these will yield the quadratic equation $x^2 - x - 1 = 0$. Solving this,

$$x = \frac{\sqrt{5}+1}{2} \quad \text{or} \quad x = \frac{-\sqrt{5}+1}{2}.$$

When

$$x = \frac{\sqrt{5}+1}{2}, \qquad y = \frac{\sqrt{5}-1}{2};$$

and when

$$x = \frac{-\sqrt{5}+1}{2}, \qquad y = \frac{-\sqrt{5}-1}{2}.$$

The first two values of x and y are ϕ and ϕ'. The second two values are their additive inverses, $-\phi$ and $-\phi'$.

Extension

To find the powers of ϕ, students first find ϕ^2 and $\phi + 1$:

$$\phi^2 = \left(\frac{\sqrt{5}+1}{2}\right)^2 = \frac{5+2\sqrt{5}+1}{4} = \frac{3+\sqrt{5}}{2},$$

$$\phi + 1 = \frac{\sqrt{5}+1}{2} + 1 = \frac{3+\sqrt{5}}{2}.$$

Then

$$\begin{aligned}\phi^3 &= \phi^2 \cdot \phi = (\phi + 1)\phi = \phi^2 + \phi \\ &= \phi + 1 + \phi = 2\phi + 1.\end{aligned}$$

The other powers of ϕ to ϕ^8 can be generated in a similar manner. They are

$$\begin{aligned}\phi^1 &= 1\phi + 0, & \phi^5 &= 5\phi + 3, \\ \phi^2 &= 1\phi + 1, & \phi^6 &= 8\phi + 5, \\ \phi^3 &= 2\phi + 1, & \phi^7 &= 13\phi + 8, \\ \phi^4 &= 3\phi + 2, & \phi^8 &= 21\phi + 13.\end{aligned}$$

The coefficients and the constants form Fibonacci sequences.

The Golden Triangle

The golden rectangle is considered to be the rectangle with the most pleasing shape. Do you remember the ratio of length to width in a golden rectangle? It's called ϕ, the golden ratio, and equals $\frac{\sqrt{5}+1}{2}$. This ratio determines whether a rectangle is "golden." In the isosceles triangle

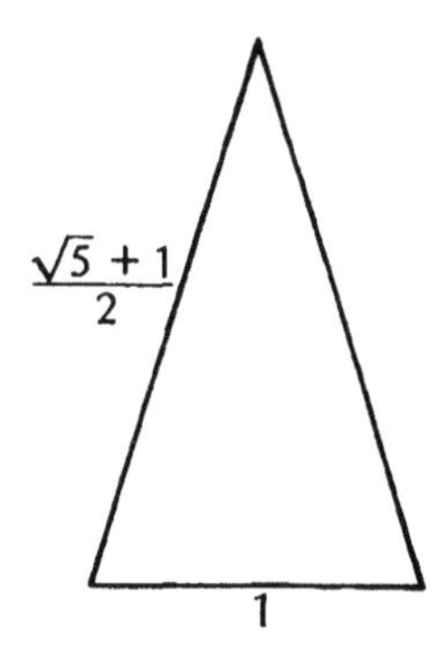

the ratio of leg to base is ϕ. As you might guess, this isosceles triangle is called a *golden triangle*. Although it isn't necessarily the triangle with the most pleasing shape, it exhibits the golden ratio.

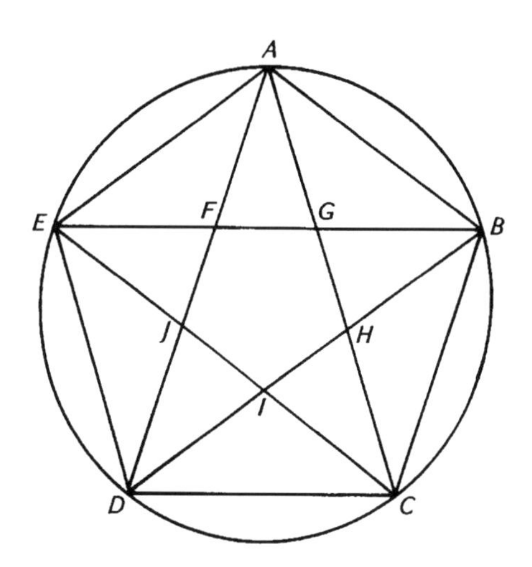

In the preceding figure, a regular pentagon is inscribed in a circle and a regular pentagram (star) is drawn. Triangle ADC is one of many golden triangles in this figure. We can use $\triangle ADC$ to find the measures of the angles of a golden triangle. The measures of arcs AB, BC, CD, DE, and EA are all equal. What is the measure of arc AB? ____________ Of arc BC? ____________ Of arc ABC? ____________ Is the measure of arc AED the same as the measure of arc ABC? ____________ Why or why not? ________________________

____

What is the measure of $\angle ADC$? ____________ Of $\angle ACD$? ____________

Why? ____

What is the measure of $\angle DAC$? ____________ Why? ________________

____

What are the measures of the angles of a golden triangle? _______________

Recall that a new golden rectangle can be constructed from a given golden rectangle. A new golden triangle can also be constructed from a given golden triangle. Look at the following figure

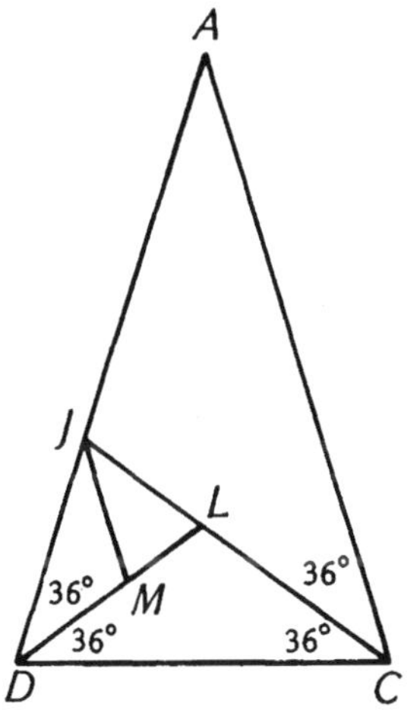

A

J

L

36°

M

36°

36°

36°

D

C

How was $\triangle CJD$ constructed? ________________________________

Is it a golden triangle? ____________ Why or why not? _______________

____

Is $\triangle DLJ$ a golden triangle? ____________ How long could you continue this process? ____

EXTENSION! Use a protractor or Geometer's Sketchpad Program to draw a golden triangle. Use a pair of compasses to construct smaller and smaller golden triangles as described. Using J as the center and AJ as the radius, draw arc AC. Continue drawing arcs to construct a golden spiral.

Teachers Notes for The Golden Triangle

This activity is a natural extension of "The Golden Rectangle," and should succeed it in a geometry course. It gives students an opportunity to apply their knowledge of angle measure in circles and triangles in a context not usually presented in geometry texts. Some of the skills and terminology presented in "The Golden Rectangle" are necessary for this activity—specifically, the golden ratio and golden spiral should be reviewed.

NCTM Standards

1	2	3	4	5	6	7	8	9	10
•	•	•	•		•	•	•		•

Presenting the Activity

Begin by briefly reviewing ϕ, the golden ratio determined in "The Golden Rectangle." Then consider the measures of the angles and arcs in the inscribed pentagon. Arcs AB, BC, CD, DE, and EA each have measure equal to $\frac{1}{5} \cdot 360°$ or $72°$. Thus, the measure of arc ABC equals $2 \cdot 72°$ or $144°$. Similarly, the measure of arc AED is $144°$. The measure of inscribed $\angle ADC$ is one-half the measure of its intercepted arc: $m\angle ADC = \frac{1}{2} \cdot 144° = 72°$. Similarly, $m\angle ACD = 72°$. The measure of $\angle DAC$ can be found in two ways:

$$\begin{aligned} m\angle DAC &= 180° - (2 \cdot 72°) = 36° \\ &= \frac{1}{2} m\overset{\frown}{DC} = \frac{1}{2} \cdot 72° = 36°. \end{aligned}$$

Thus, the measures of the angles of a golden triangle are $36°$, $72°$, and $72°$.

Note that a proof that $\triangle ADC$ is a golden triangle is not presented on the student page. If time permits, you may want to present it in class as follows. Assume each side of pentagon $ABCDE$ has length ϕ. Using angle measure in a circle, we can find the angle measures shown on the following figure.

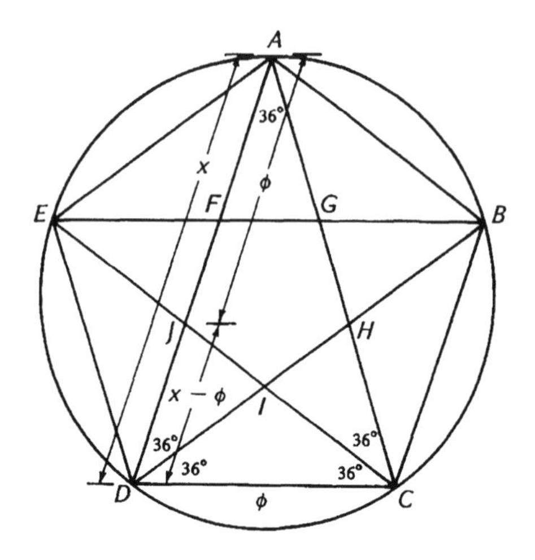

We can see that $\triangle ADC$ and $\triangle CDJ$ are similar isosceles triangles. Let $AD = AC = x$. Then $DJ = x - \phi$, and because $DC = \phi$,

$$\frac{DC}{DJ} = \frac{AC}{DC} \quad \text{or} \quad \frac{\phi}{x - \phi} = \frac{x}{\phi}.$$

Thus, $x^2 - \phi x - \phi^2 = 0$. The positive root of this equation from the quadratic formula is

$$x = \phi\left(\frac{1 + \sqrt{5}}{2}\right),$$

but by definition,

$$\frac{1 + \sqrt{5}}{2} = \phi.$$

Hence, $x = \phi \cdot \phi = \phi^2$, and because $AD = x$, $AD = \phi^2$. Thus, in $\triangle ADC$, the ratio of leg to base is

$$\frac{AD}{DC} = \frac{\phi^2}{\phi} = \phi.$$

Also, $DJ = x - \phi = \phi^2 - \phi = \phi + 1 - \phi = 1$, so in $\triangle CDJ$, the ratio of leg to base is

$$\frac{CD}{DJ} = \frac{\phi}{1} = \phi.$$

Therefore, in any 72°–72°–36° isosceles triangle, the ratio of leg to base is ϕ and any triangle with these angle measures is a golden triangle.

The next paragraph explores how a new golden triangle can be constructed from a given golden triangle. $\triangle CDJ$ is constructed by bisecting $\angle C$. It is a golden triangle because its angles measure 36°, 72°, and 72°. $\triangle DLJ$ is also a golden triangle. This process can be continued indefinitely to produce a series of smaller and smaller golden triangles.

Extension

The figure constructed by the students should be similar to the following one:

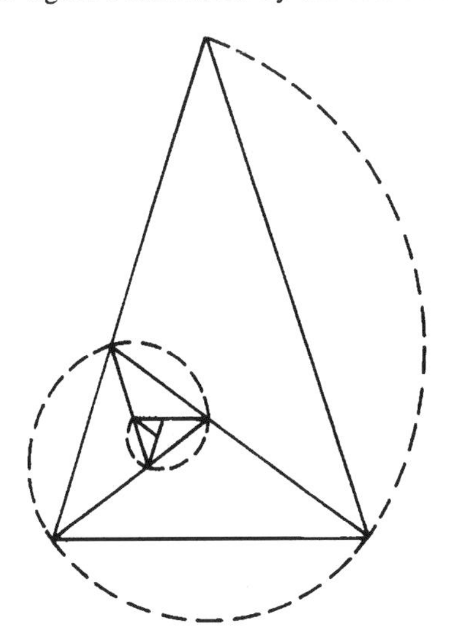

The Arbelos

Archimedes (287–212 B.C.) is considered to be one of the greatest mathematicians of all time. He was killed by a Roman soldier who found him drawing geometric figures in the sand. According to one source, Archimedes' last words were, "Don't disturb my circles!" One figure Archimedes studied is

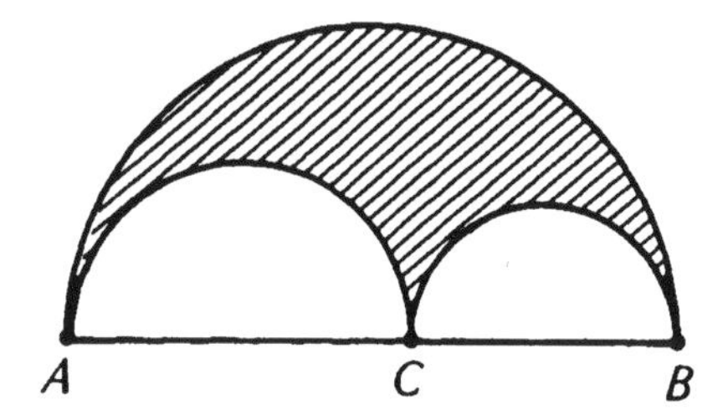

The shaded portion is known as the *arbelos or shoemaker's knife*. One property of the arbelos is that for any point C on $\overline{AB}$, the semicircle on $\overline{AC}$ plus the semicircle on $\overline{BC}$ equals the semicircle on $\overline{AB}$. We can prove this by finding the lengths of the semicircular arcs:

Let the radii of semicircles on $\overline{AC}$, $\overline{BC}$, and $\overline{AB}$ be r_1, r_2, and R, respectively. Thus

$$\ell\overset{\frown}{AC} = \frac{1}{2} \cdot 2\pi r_1 = \pi r_1,$$

$$\ell\overset{\frown}{BC} = \underline{\hspace{4cm}},$$

$$\ell\overset{\frown}{AB} = \underline{\hspace{4cm}}.$$

We know that $R = r_1 + r_2$, so $\pi R =$ ________________________.

Therefore, $\ell\overset{\frown}{AB} =$ ________________________.

Suppose three (or more) semicircles are drawn on $\overline{AB}$ as shown:

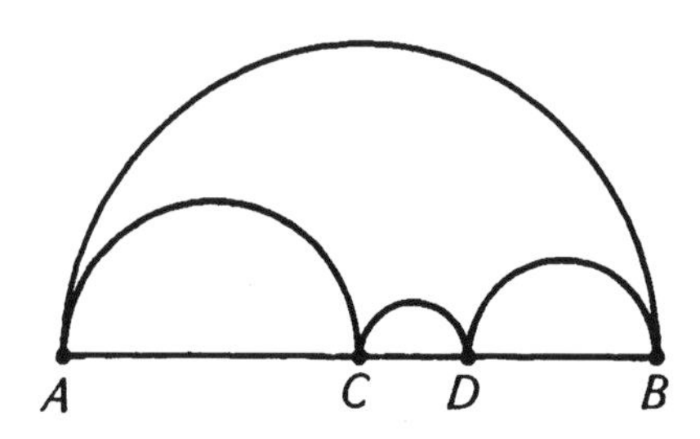

What relationship do you think would exist among these semicircular arcs? ___

____

Now consider some other properties of the arbelos. In the following figure, $\overline{HC} \perp \overline{ACB}$ at C, $\overline{FG}$ is a common external tangent of the two smaller semicircles, and $\overline{JE} \perp \overline{FD}$.

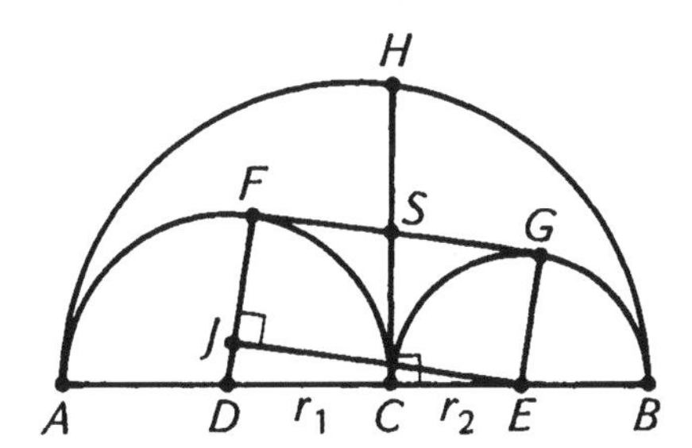

Because line segment drawn perpendicular to a diameter is the geometric mean between the segments of the diameter, $\frac{2r_1}{HC} = \frac{HC}{2r_2}$. Thus, $(HC)^2 =$ __________.
Also, $FG = JE$. Why? ____
Express JD and DE in terms of r_1 and r_2:

$JD =$ __________________ $DE =$ __________________.

The use the Pythagorean theorem to find $(JE)^2$:

$(JE)^2 =$ __________________.

Because $JE = FG$, $(JE)^2 = (FG)^2$. Thus, $(FG)^2 =$ __________________.
What does this tell about the relationship between HC and FG? __________
Now, verify that $\overline{HC}$ and $\overline{FG}$ bisect each other at S. ____________________

____

Do the points F, H, G, C determine a circle with center S? ______________
Why or why not? ____

EXTENSION! Prove that the area of the arbelos equals the area of the circle with diameter $\overline{HC}$.

Teacher's Notes for The Arbelos

Many of the problems studied by the ancient geometers are not included in most standard geometry texts. Unfortunately, the arbelos is one of these. The arbelos provides students with an opportunity to apply what they know about several areas of geometry—properties of triangles, rectangles, and circles are all needed to complete the activity. Students will find that one interesting relationship leads to another, and then another—all derived from a relatively simple drawing of semicircles.

This activity should usually be presented later in a geometry course because students must be able to compute lengths of arcs and areas of circles.

NCTM Standards

1	2	3	4	5	6	7	8	9	10
●	●	●	●		●	●		●	●

Presenting the Activity

Students can use the formula for the length of an arc: $\frac{n}{360} \times 2\pi r$, where n is the number of degrees in the arc and r is the length of the radius of the circle. Alternatively, they can simply use the fact that a semicircle is half the circumference of the circle. Thus, $\ell\overset{\frown}{AC} = \pi r_1$, $\ell\overset{\frown}{BC} = \pi r_2$, and $\ell\overset{\frown}{AB} = \pi R$. Whereas $R = r_1 + r_2$, $\pi R = \pi r_1 + \pi r_2$ and $\ell\overset{\frown}{AB} = \ell\overset{\frown}{AC} + \ell\overset{\frown}{BC}$. For the three semicircles on the student page, $\ell\overset{\frown}{AB} = \ell\overset{\frown}{AC} + \ell\overset{\frown}{CD} + \ell\overset{\frown}{DB}$. If students question this relationship, have them verify it using the same procedure as used for two semicircles.

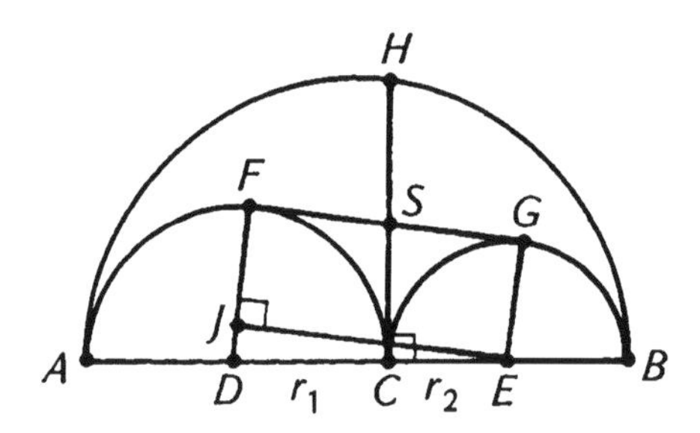

The next series of questions leads students to the discovery that $HC = FG$. They begin by finding that $(HC)^2 = 4r_1r_2$. To show that $FG = JE$, students must show that $FGEJ$ is a rectangle. Whereas $\overline{DF} \perp \overline{FG}$ and $\overline{EG} \perp \overline{FG}$ (a radius is perpendicular to a tangent at the point of tangency), and it is given that $\overline{JE} \perp \overline{FD}$, $FGEJ$ has four right angles. Therefore, $FGEJ$ is a rectangle. Expressing JD and DE in terms of r_1 and r_2, we have $JD = r_1 - r_2$ and $DE = r_1 + r_2$. Then, in $\triangle JDE$, we have

$$\begin{aligned}(JE)^2 &= (DE)^2 - (JD)^2 \\ &= (r_1 + r_2)^2 - (r_1 - r_2)^2 \\ &= r_1^2 + 2r_1r_2 + r_2^2 - r_1^2 + 2r_1r_2 - r_2^2 = 4r_1r_2\end{aligned}$$

Thus, $(FG)^2 = 4r_1r_2$ and $HC + FG$.

Now students can verify that $\overline{HC}$ and $\overline{FG}$ bisect each other at S: $\overline{SC}$ is a common internal tangent to both circles. Therefore, $FS = SC$ and $SC = SG$, which gives us

$FS = SG$. However, because $HC = FG$, $HC = HS + SC$, and $FG = SC + SC$, $HC + SC = SC + SC$. Therefore, we also know that $HS = SC$, so $\overline{HC}$ and $\overline{FG}$ bisect each other at S.

The points F, H, G, C determine a circle with center S because $FS = GS = HS = SC$.

If time permits, you may want to give students values for r_1 and r_2 such as $r_1 = 16$ and $r_2 = 4$. Then have them use these values to show that $\ell\widehat{AB} = \ell\widehat{AC} + \ell\widehat{BC}$ and to find the radius of circle S (8) and the area of the arbelos. (After completing the Extension, they would find the area is 64π.)

Extension

Express the area of the arbelos in terms of r_1 and r_2. The area of the arbelos is equal to the area of semicircle AHB minus (the area of semicircle AFC plus the area of semicircle CGB).

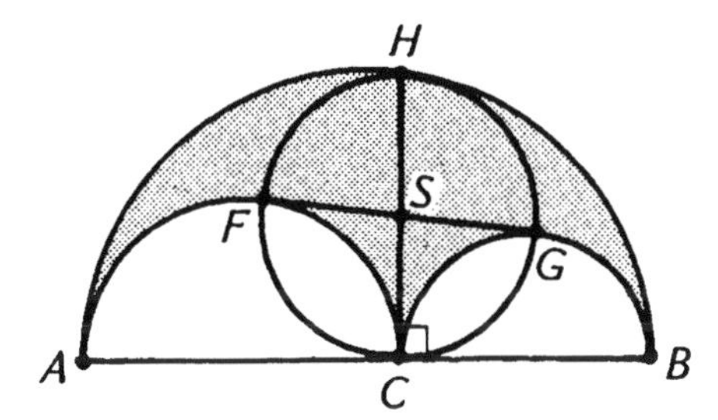

Whereas the area of semicircle is equal to $\frac{\pi r^2}{2}$, we have

$$\begin{aligned}\text{area of arbelos} &= \frac{\pi R^2}{2} - \frac{\pi r_1^2}{2} - \frac{\pi r_2^2}{2}\\ &= \frac{\pi}{2}(R^2 - r_1^2 - r_2^2).\end{aligned}$$

We know that $R = r_1 + r_2$. Substituting, we get

$$\begin{aligned}\text{area of arbelos} &= \frac{\pi}{2}((r_1 + r_2)^2 - r_1^2 - r_2^2)\\ &= \frac{\pi}{2}(r_1^2 + 2r_1r_2 + r_2^2 - r_1^2 - r_2^2)\\ &= \frac{\pi}{2}(2r_1r_2)\\ &= \pi r_1 r_2.\end{aligned}$$

Now find the area of circle S. We found earlier that $(HC)^2 = 4r_1r_2$, so the diameter $HC = 2\sqrt{r_1r_2}$. Therefore, the radius of circle S is $\sqrt{r_1r_2}$. Then the area of the circle equals $\pi(\sqrt{r_1r_2})^2 = \pi r_1 r_2$. Thus, the area of the arbelos is equal to the area of circle S.

There are many other fascinating properties and extensions of the arbelos. Some are included in Martin Gardner's "Mathematical Games" in *Scientific American* (Vol. 240, No. 1, pp. 18–28, January 1979).

CHAPTER 3

Post-Euclidean Theorems

- Ptolemy's Theorem
- Ceva's Theorem
- Stewart's Theorem
- Simson's Theorem ♦
- Napoleon's Theorem ★

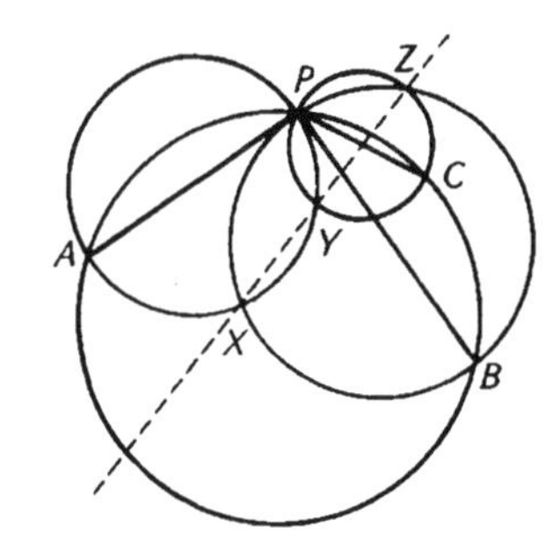

Most students know that geometry originated with the ancient Greeks and was organized into logical form by Euclid. However, most students don't realize how much geometry has changed and grown since Euclid's time. Many theorems of Euclidean geometry were developed and proved hundreds of years later. Five of these theorems are shown in this section.

Many of the new theorems based on Euclidean geometry were discovered in the scientific boom that began during the Renaissance. All the activities in this category except "Ptolemy's Theorem" explore theorems developed during this period.

"Ptolemy's Theorem" uses cyclic quadrilaterals to develop a method for finding the length of the diagonals of a quadrilateral. Surprisingly, students need not be familiar with properties of circles. Only the idea of an inscribed polygon is necessary, and a brief discussion in class of this concept should be sufficient. Thus, "Ptolemy's Theorem" is an excellent way to use circles without needing any properties of circles.

Many texts do not prove that the medians of a triangle are concurrent; others prove it very late in the course and must introduce coordinate geometry first. "Ceva's theorem" presents a very simple method for proving the concurrency theorems for medians, angle bisectors, and altitudes. Only a knowledge of similar triangles and the angle bisector theorem is required. The proof of Ceva's theorem, given in the Teacher's Notes, has two parts and two cases. Thus, it may be best to present it only to your better students.

In "Ceva's Theorem," students prove that certain cevians—segments joining a vertex of a triangle and a point on the opposite side—are concurrent. In "Stewart's Theorem," students learn a formula for finding the length of *any* cevian of a triangle. The proof of Stewart's theorem is given in the Teacher's Notes and provides a good algebra review and excellent reinforcement for the Pythagorean theorem. Stewart's theorem shows the importance of the development of algebra to the growth of geometry.

Problems and theorems about collinear points are seldom included in geometry texts because proving that points are collinear is often difficult. The proofs in "Simson's Theorem" are an exception. Although students will probably not be able to write the proof of Simson's theorem or the proof in the Extension, they should be able to understand them if they are discussed in class.

"Napoleon's Theorem" is interesting not only because of the fame of its discoverer, but because, as with "The Arbelos," it tests many areas of geometric knowledge: congruence and similarity in triangles, the ratios in 30°-60°-90° triangles, and angle measure in circles. "Minimizing Distances," an application using the equiangular or Napoleon point, can immediately follow this activity.

Ptolemy's Theorem

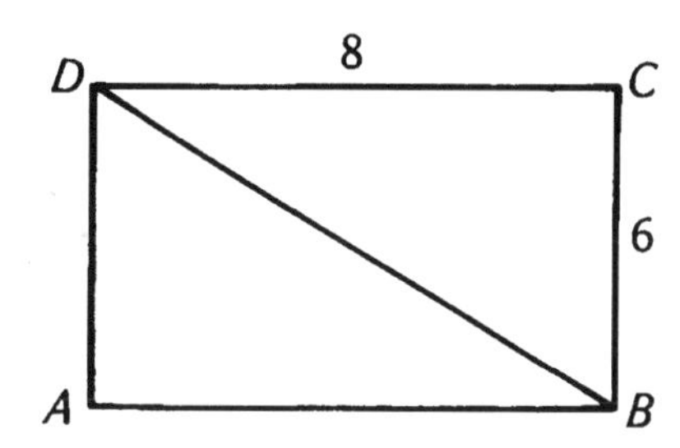

Can you find the length of diagonal $\overline{BD}$ of rectangle $\overline{ABCD}$ in the preceding diagram? What theorem should you use? ________________________________

What is the length of $\overline{BD}$? ____________

Suppose we now consider finding the length of diagonal $\overline{HF}$ of the isosceles trapezoid $EFGH$:

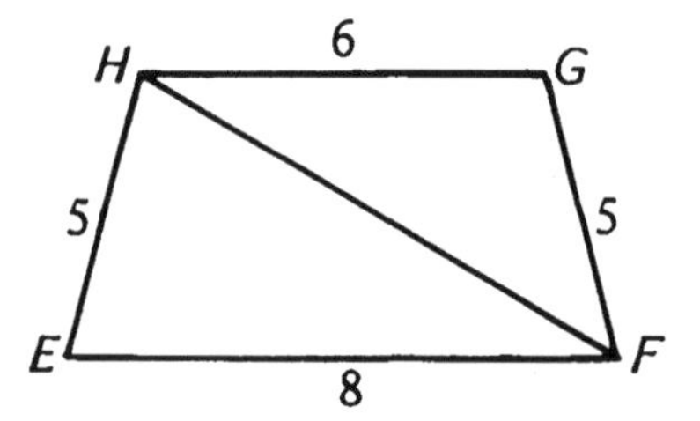

Can you use the same theorem you used to find the length of $\overline{HF}$? _________

This problem was studied by Claudius Ptolemy, who lived in Alexandria about 150 A.D. The following theorem, known as Ptolemy's theorem, was included in his book, *The Almagest*:

> *In a cyclic (inscribed) quadrilateral, the product of the lengths of the diagonals is equal to the sum of the products of the lengths of the pairs of opposite sides.*

Look at the following figure:

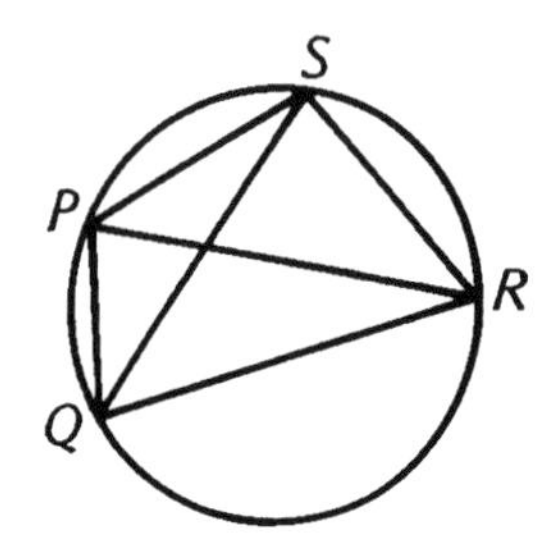

Write the relationship given in Ptolemy's theorem as an equation using the quadrilateral in the figure. ____

Ptolemy's theorem can help find the length of a diagonal of a quadrilateral only if we know the quadrilateral is cyclic. It is possible to prove that a quadrilateral is cyclic if and only if its opposite angles are supplementary. Is any trapezoid cyclic? ____________ Is any isosceles trapezoid cyclic? ____________

So, we can use Ptolemy's theorem to find the length of $\overline{HF}$ in the previous trapezoid $EFGH$. What is the length of $\overline{HF}$? ____________

Can we apply Ptolemy's theorem to a rectangle? That is, is a rectangle cyclic? ____________ Try it. What theorem do you get? ____________________

Now apply Ptolemy's theorem to this problem: Point P on the accompanying diagram is on arc AB of the circumscribed circle of equilateral $\triangle ABC$, $AP = 3$, and $BP = 4$. What is the length of $\overline{CP}$? __________________________

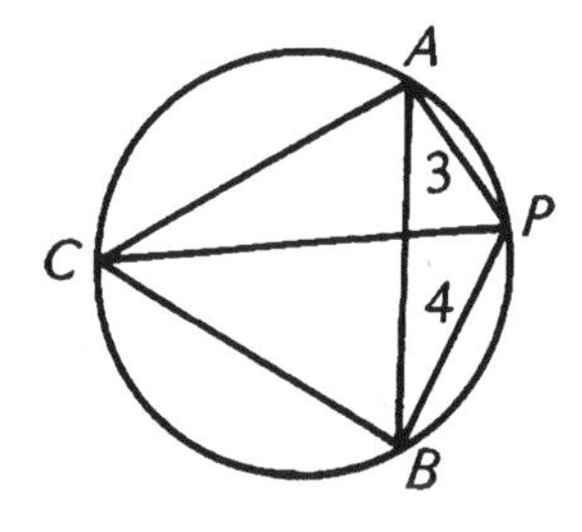

EXTENSION! Using the following figure, where $\angle BAC \cong \angle DAP$, write a proof of Ptolemy's theorem.

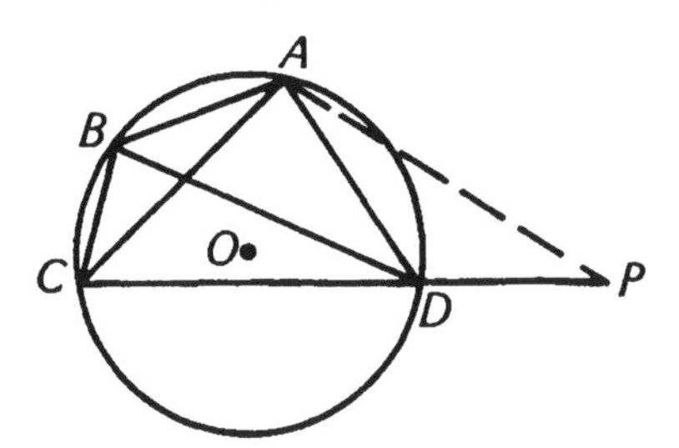

Teacher's Notes for Ptolemy's Theorem

Students often encounter problems and proofs that appear to be very difficult—until they find the right theorem to use. This activity provides students with just such a theorem concerning cyclic quadrilaterals. They will discover when to apply Ptolemy's theorem and how it relates to other concepts in geometry.

For everything except the Extension, students only need to be familiar with the basic properties of regular polygons, rectangles, and trapezoids and with the Pythagorean theorem. However, to write the proof given in the Extension, students will need to know how to prove that two triangles are similar.

NCTM Standards

1	2	3	4	5	6	7	8	9	10
	•	•	•		•	•		•	•

Presenting the Activity

By using the Pythagorean theorem, students will find that the length of diagonal $\overline{BD}$ of rectangle $ABCD$ is 10. When students consider the isosceles trapezoid, they may make too quick a decision and assume they can't use the Pythagorean theorem to find HF. Point out that by drawing a perpendicular from H to $\overline{EF}$ and applying the Pythagorean theorem twice, they can find the length of $\overline{HF}$. However, most students, after being shown this method, will welcome a less tedious one.

Take some time to consider the statement of Ptolemy's theorem, making sure students can write the relationship as an equation using quadrilateral $PQRS$: $(PR)(QS) = (PS)(QR) + (PQ)(SR)$. Then consider when Ptolemy's theorem can be used. Ask students which quadrilaterals have opposite angles that are supplementary. They should realize that all squares, rectangles, and isosceles trapezoids have opposite angles that are supplementary and are therefore cyclic quadrilaterals.

Now they can apply Ptolemy's theorem to trapezoid $EFGH$: $(HF)(EG) = (HG)(EF) + (HE)(GF)$. Because $HF = EG$, $(HF)^2 = (6)(8) + (5)(5) = 73$ and $HF = \sqrt{73}$.

When students apply Ptolemy's theorem to a rectangle, they should be surprised. They will find that

$$(WY)(ZX) = (ZY)(WX) + (WZ)(XY)$$

for a rectangle such as $WXYZ$:

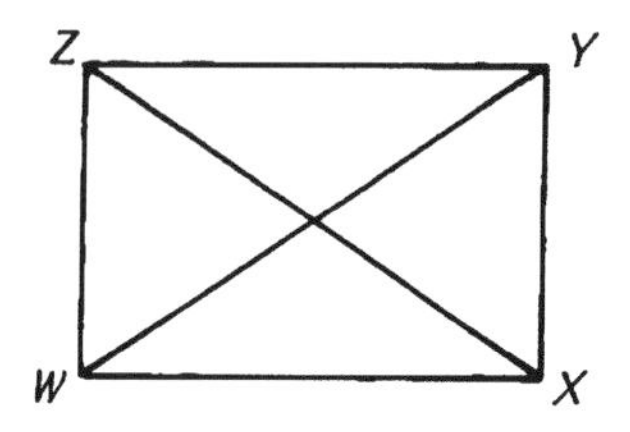

However, $ZY = WX$, $XY = WZ$, and $WY = ZX$. Therefore, by substitution, $(ZX)^2 = (WX)^2 + (WZ)^2$. This, of course, is the Pythagorean theorem.

In the next problem, let t represent the length of a side of equilateral $\triangle ABC$. Because quadrilateral $APBC$ is cyclic, we may apply Ptolemy's theorem:

$$(CP)(t) = (AP)(t) + (BP)(t).$$

Therefore, $CP = AP + BP = 3 + 4 = 7$.

You may wish to have students try to generalize the concept illustrated by this problem: Given any equilateral triangle that is inscribed in a circle and any point on the circle that is not a vertex of the triangle.

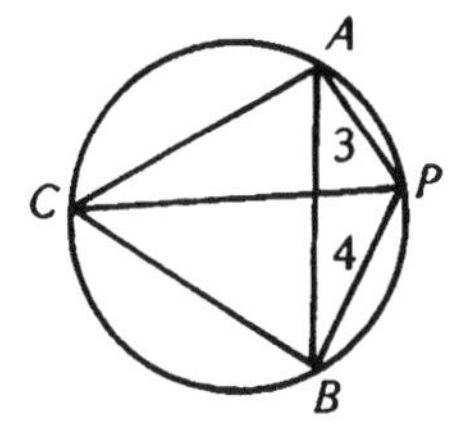

The sum of the distances from the point to the closer vertices equals the distance from the point to the further vertex.

Students should be encouraged to investigate similar problems where the equilateral triangle is replaced with other regular polygons.

Extension

The proof of Ptolemy's theorem is based on similar triangles:

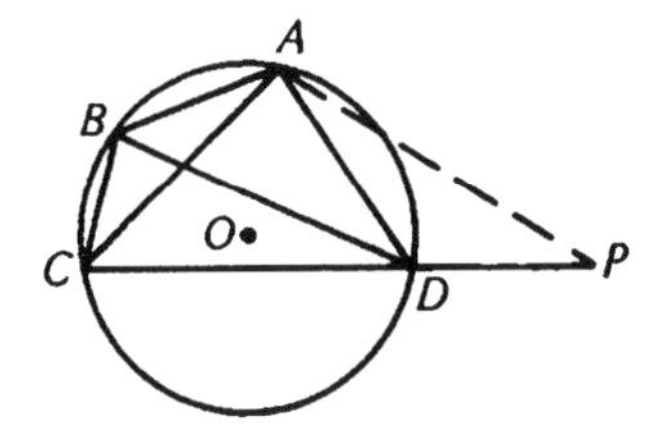

Because quadrilateral $ABCD$ is cyclic, $\angle ABC$ is supplementary to $\angle ADC$. However, $\angle ADP$ is also supplementary to $\angle ADC$. Therefore, $m\angle ABC = m\angle ADP$. Whereas $\overline{AP}$ was drawn so that $\angle BAC \cong \angle DAP$, we know that $\triangle BAC \sim \triangle DAP$. Thus,

$$\frac{AB}{AD} = \frac{BC}{DP} \quad \text{or} \quad DP = \frac{(AD)(BC)}{AB}.$$

Whereas $m\angle BAC = m\angle DAP$, $m\angle BAD = m\angle CAP$, and whereas $\triangle BAC \sim \triangle DAP$, $\frac{AB}{AD} = \frac{AC}{AP}$. Therefore, $\triangle ABD \sim \triangle ACP$. Then

$$\frac{BD}{CP} = \frac{AB}{AC} \quad \text{or} \quad CP = \frac{(AC)(BD)}{AB},$$

but $CP = CD + DP$. So, by substitution,

$$\frac{(AC)(BD)}{AB} = CD + \frac{(AD)(BC)}{AB}.$$

Now simplifying this expression gives us the desired result:

$$(AC)(BD) = (AB)(CD) + (AD)(BC).$$

Ceva's Theorem

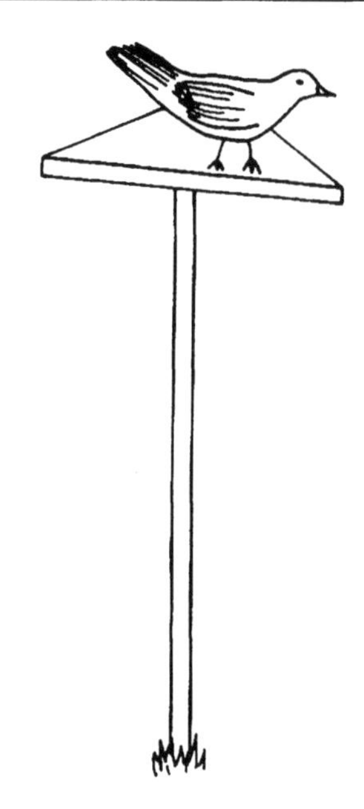

Have you ever tried to balance a ruler on your finger? The point where it balances is its *center of gravity*. If the ruler has uniform thickness and density, where do you think its center of gravity will be? ____________________

Now suppose you have a triangular board that you want to use as a platform for a bird feeder. Why would it help to know where the board's center of gravity is?

____

The center of gravity of a triangle is called the *centroid*. It's the point where the medians of the triangle intersect. Can you prove that the medians of a triangle are concurrent (intersect in a point)? Try it.

It's easy to prove the medians of a triangle are concurrent if you know a theorem published in 1678 by the Italian mathematician Giovanni Ceva. This theorem states:

> Three lines drawn from the vertices A, B, and C of $\triangle ABC$ meeting the opposite sides in points L, M, and N, respectively, are concurrent if and only if
>
> $$\frac{AN}{NB} \cdot \frac{BL}{LC} \cdot \frac{CM}{MA} = 1.$$

Now let's prove that the medians $\overline{AL}$, $\overline{BM}$, and $\overline{CN}$ of the following $\triangle ABC$ are concurrent.

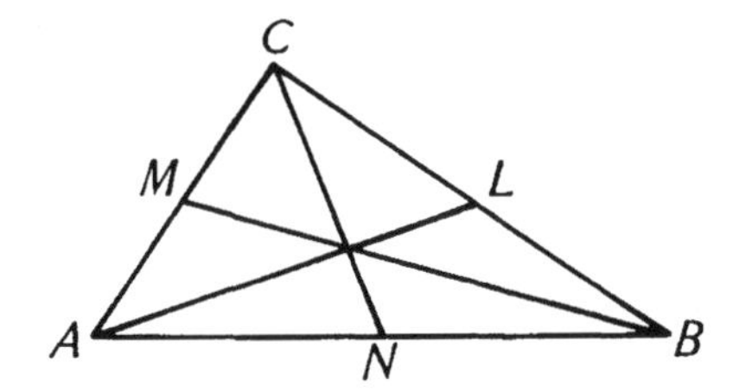

We know that $AN = NB$, $BL = LC$, and $CM = MA$. Why? ____________

____

By multiplication, $(AN)(BL)(CM) =$ ______________________________

and $\frac{AN}{NB} \cdot \frac{BL}{LC} \cdot \frac{CM}{MA} =$ ______________________________________.

Does this prove the medians are concurrent? Why or why not? ____________

____

Now consider the three interior angle bisectors of $\triangle ABC$

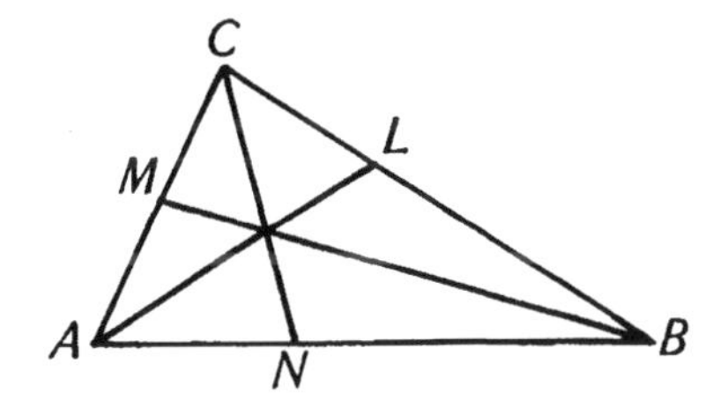

Are they concurrent? Because $\overline{AL}$ bisects $\angle CAB$, $\frac{BL}{LC} = \frac{AB}{AC}$. Why? ________

____

Similarly, $\frac{AN}{NB} =$ ____________ and $\frac{CM}{MA} =$ ____________.

By multiplication, $\frac{AN}{NB} \cdot \frac{BL}{LC} \cdot \frac{CM}{MA} =$ _______________________________.

Does this prove the interior angle bisectors are concurrent? ____________

What special circle can be drawn using this point of concurrency? __________

____

EXTENSION! Using in the accompanying diagram, $\triangle ABC$ prove that the altitudes $\overline{AL}$, $\overline{BM}$, and $\overline{CN}$ are concurrent. What difficulty do you have if you try to apply this proof to a right triangle?

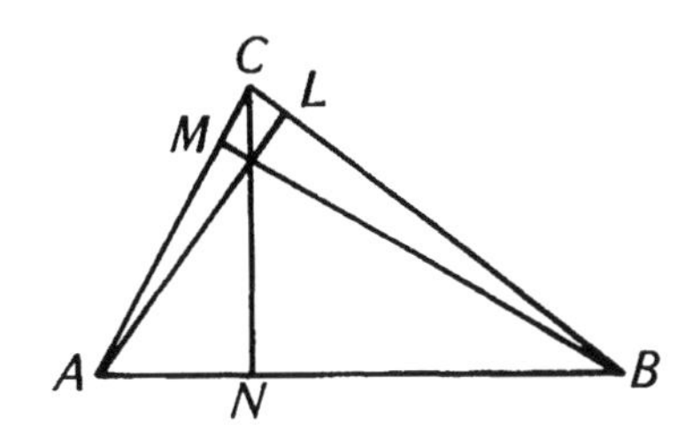

Teacher's Notes for Ceva's Theorem

Most of the theorems studied in high school geometry were considered by Euclid. Unfortunately, this can give students the idea that nothing has been discovered since his time. We know, of course, that this isn't the case. Ceva's theorem for proving lines concurrent was one of several that appeared in the seventeenth century. This activity presents Ceva's theorem, giving students an opportunity to shorten proofs that deal with concurrent lines.

The activity should be presented after students have studied similar triangles and the angle-bisector theorem. They should also be familiar with the definitions of median, altitude, and inscribed circle of a triangle.

NCTM Standards

1	2	3	4	5	6	7	8	9	10
	•	•	•		•	•		•	•

Presenting the Activity

By experimenting, students will find that if a ruler has uniform thickness and density, its center of gravity is at the midpoint of its length. The bird feeder will be stable if the post supporting the feeder is attached at the triangular board's center of gravity.

It is possible to prove that the medians of a triangle are concurrent without using Ceva's theorem. However, most students would be unlikely to think of how to do so.

A proof of Ceva's theorem follows. Note that it is biconditional; thus, there are two parts. Also, there are two cases, as shown by the figures below.

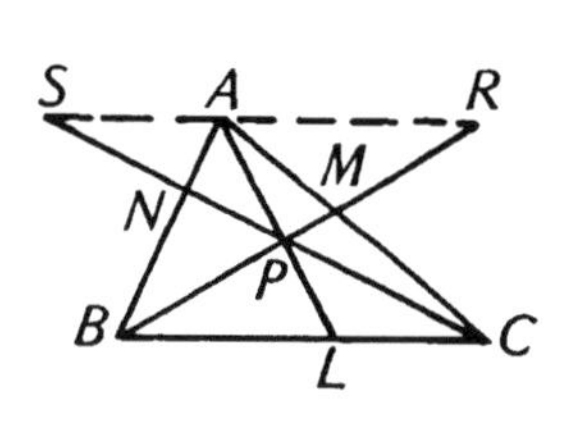

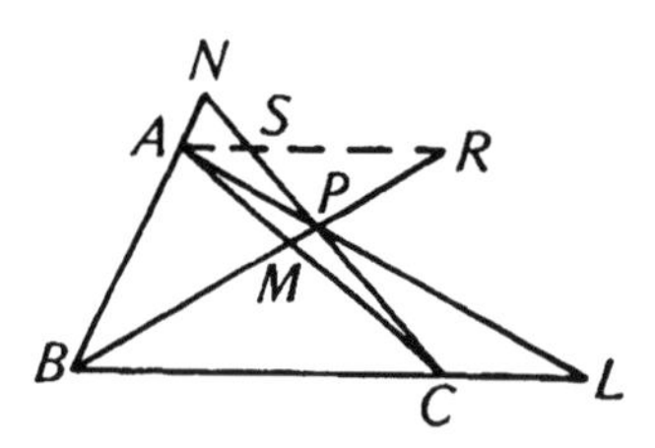

The three lines may meet inside or outside the given triangle. For part 1, we are given $\triangle ABC$, with N on $\overleftrightarrow{AB}$, M on $\overleftrightarrow{AC}$, and L on $\overleftrightarrow{BC}$; also $\overleftrightarrow{AL}$, $\overleftrightarrow{BM}$, and $\overleftrightarrow{CN}$ are concurrent at P. We want to prove that

$$\frac{AN}{NB} \cdot \frac{BL}{LC} \cdot \frac{CM}{MA} = 1.$$

Draw a line through A, parallel to $\overleftrightarrow{BC}$ meeting $\overleftrightarrow{CP}$ at S and $\overleftrightarrow{BP}$ at R:

$$\triangle CLP \sim \triangle SAP, \quad \text{so } \frac{CL}{SA} = \frac{LP}{AP};$$

$$\triangle BLP \sim \triangle RAP, \quad \text{so } \frac{BL}{RA} = \frac{LP}{AP}.$$

Therefore,

$$\frac{CL}{SA} = \frac{BL}{RA}, \quad \text{or} \quad \frac{CL}{BL} = \frac{SA}{RA};$$

$$\triangle AMR \sim \triangle CMB, \quad \text{so } \frac{AM}{MC} = \frac{AR}{CB};$$

$$\triangle BNC \sim \triangle ANS, \quad \text{so } \frac{BN}{NA} = \frac{CB}{SA}.$$

Multiplying, we have

$$\frac{AM}{MC} \cdot \frac{BN}{NA} \cdot \frac{CL}{BL} = \frac{AR}{CB} \cdot \frac{CB}{SA} \cdot \frac{SA}{RA} = 1.$$

For part 2, let $\overleftrightarrow{BM}$ and $\overleftrightarrow{AL}$ meet at P. Let $\overleftrightarrow{CP}$ meet $\overleftrightarrow{AB}$ at N', so $\overleftrightarrow{AL}$, $\overleftrightarrow{BM}$ and $\overleftrightarrow{CN'}$ are concurrent at P. By the part of Ceva's theorem we have already proved, we get:

$$\frac{BL}{LC} \cdot \frac{CM}{MA} \cdot \frac{AN'}{N'B} = 1.$$

However, it is given that

$$\frac{BL}{LC} \cdot \frac{CM}{MA} \cdot \frac{AN}{NB} = 1.$$

Therefore,

$$\frac{AN'}{N'B} = \frac{AN}{NB},$$

so that N and N' must coincide. Thus, the three lines are concurrent.

The proof that the medians are concurrent is unambiguous using Ceva's theorem. The lengths of the segments are equal because L, M, and N are midpoints by the definition of a median. Multiplying gives $(AN)(BL)(CM) = (NB)(LC)(MA)$ or

$$\frac{AN}{NB} \cdot \frac{BL}{LC} \cdot \frac{CM}{MA} = 1.$$

By Ceva's theorem then, the lines are concurrent.

To show that the angle bisectors of a triangle are concurrent, students must recall that an angle bisector of a triangle divides the opposite side into segments proportional to the two remaining sides of the triangle. It follows that

$$\frac{AN}{NB} = \frac{AC}{BC},$$

$$\frac{BL}{LC} = \frac{AB}{AC},$$

and

$$\frac{CM}{MA} = \frac{BC}{AB}.$$

Then by multiplying

$$\frac{AN}{NB} \cdot \frac{BL}{LC} \cdot \frac{CM}{MA} = \frac{AC}{BC} \cdot \frac{AB}{AC} \cdot \frac{BC}{AB} = 1.$$

Thus, by Ceva's theorem the three angle bisectors are concurrent. The point of concurrency of the three angle bisectors is the center of the triangle's inscribed circle.

Extension

To prove the altitudes are concurrent, students must use similar triangles:

$$\triangle ANC \sim \triangle AMB, \quad \text{so } \frac{AN}{MA} = \frac{AC}{AB};$$

$$\triangle BLA \sim \triangle BNC, \quad \text{so } \frac{BL}{NB} = \frac{AB}{BC};$$

$$\triangle CMB \sim \triangle CLA, \quad \text{so } \frac{CM}{LC} = \frac{BC}{AC}.$$

By multiplying, we get

$$\frac{AN}{MA} \cdot \frac{BL}{NB} \cdot \frac{CM}{LC} = \frac{AC}{AB} \cdot \frac{AB}{BC} \cdot \frac{BC}{AC} = 1.$$

Thus, by Ceva's theorem the altitudes are concurrent.

If $\triangle ABC$ is a right triangle, the altitudes are concurrent at the right angle and the proof is obvious.

Stewart's Theorem

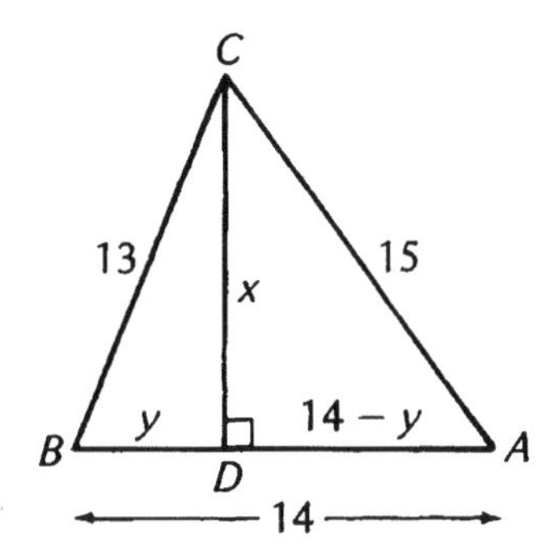

In $\triangle ABC$ in the preceding diagram, you can find x, the length of the altitude to $\overline{AB}$, by using the Pythagorean theorem twice:

$$x^2 + (14 - y)^2 = 225,$$
$$x^2 + y^2 = 169.$$

By subtraction, $(14 - y)^2 - y^2 = 56$. What is the value of y? _______ What is the value of x? _______

Suppose $\overline{CD}$ is *any* line segment joining C to *any* point D on $\overline{AB}$. Such a line segment is called a *cevian* of a triangle. It's possible to find the length of a cevian using the Pythagorean theorem, but there's a lot of computation involved. Fortunately, Matthew Stewart published a theorem in 1745 that makes it much easier to find the length of any cevian of a triangle.

Stewart's theorem: In the accompanying figure, $a^2n + b^2m = c(d^2 + mn)$.

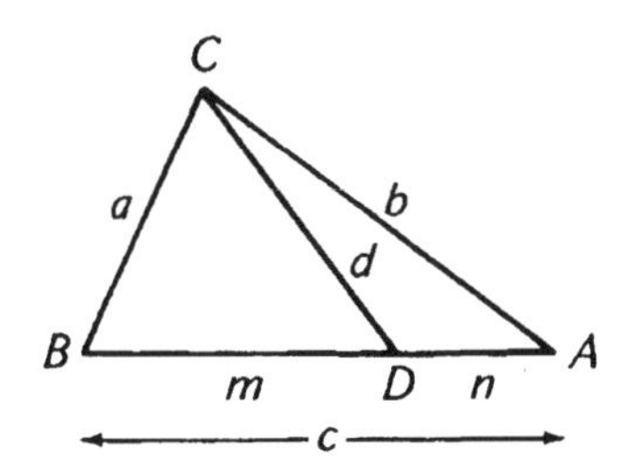

Use Stewart's theorem to find x, the length of cevian $\overline{RS}$, in following figure:

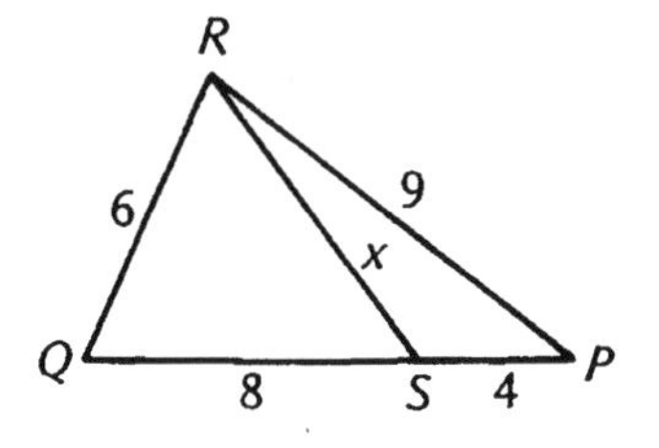

What is the value of x? _______

Altitudes are special kinds of cevians. So are medians and angle bisectors. Consider the median to $\overline{AB}$ of $\triangle ABC$:

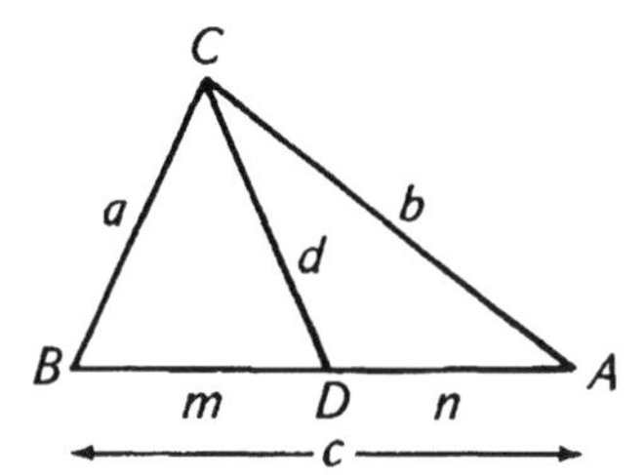

What is true about m and n? ______________________________

Write m and n in terms of c: ______________________________

Substitute these values of m and n in Stewart's theorem and solve for d^2. Now write a general equation for finding the length of a median of a triangle:

$d^2 =$ ______________________________.

Use your equation to find the length of the median to $\overline{XZ}$ in the triangle

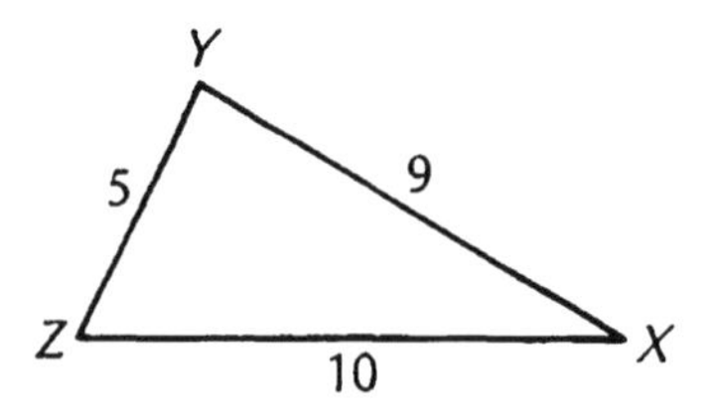

____

____

EXTENSION! In $\triangle ABC$, $\overline{CD}$ is the angle bisector of $\angle BCA$. Show that Stewart's theorem can be written as $d^2 = ab - mn$ when d is the length of the angle bisector shown in the accompanying figure. Then find d when $a = 9$, $b = 15$, $m = 6$, and $n = 10$.

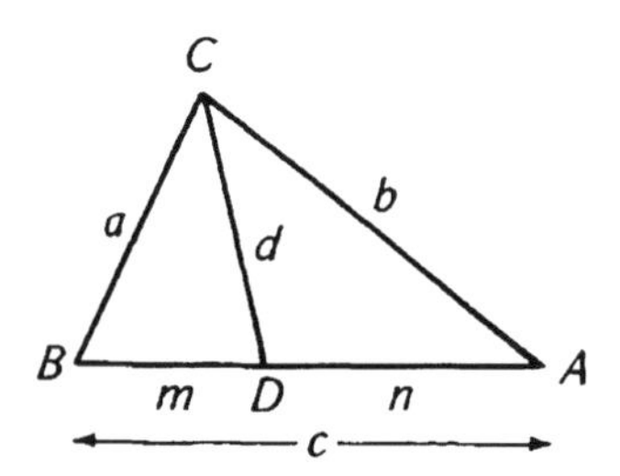

Teacher's Notes for Stewart's Theorem

Although students are familiar with the properties of altitudes, medians, and angle bisectors of triangles, finding the lengths of these segments will probably be a new experience. This activity presents a theorem that makes it easy to find the length of any segment joining a vertex of a triangle to any point on the opposite side. The activity also reinforces the Pythagorean theorem and provides an excellent algebra review.

Students will see how algebra can be used to simplify geometric problem solving.

Students must know the Pythagorean theorem and the properties of altitudes, medians, and angle bisectors of triangles.

NCTM Standards									
1	2	3	4	5	6	7	8	9	10
•	•	•	•		•	•		•	•

Presenting the Activity

Students should be able to complete the first problem easily as shown:

$$(14 - y)^2 - y^2 = 56,$$
$$196 - 28y + y^2 - y^2 = 56,$$
$$-28y = -140,$$
$$y = 5;$$
$$x^2 + y^2 = 169,$$
$$x^2 + 25 = 169,$$
$$x^2 = 144,$$
$$x = 12.$$

Cevians are named after Giovanni Ceva. His theorem about the concurrency of cevians is presented in "Ceva's Theorem." The proof of Stewart's theorem is lengthy but not difficult. It is given in the following text and you may wish to present it if time permits.

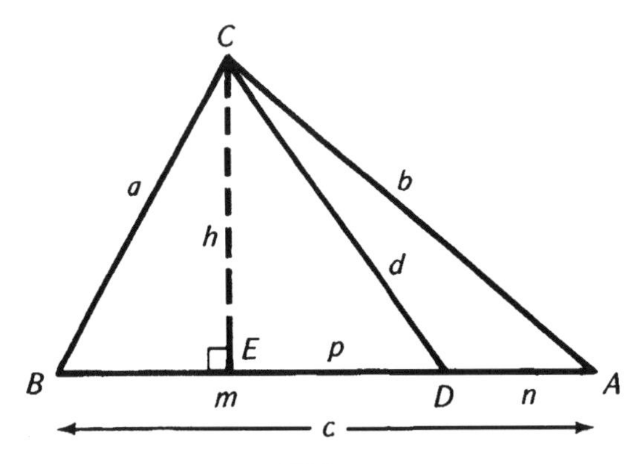

In $\triangle ABC$, draw altitude $\overline{CE}$ and let $CE = h$ and let $ED = p$. Apply the Pythagorean theorem to $\triangle CEB$ to obtain

$$a^2 = h^2 + (m - p)^2. \tag{1}$$

By applying the Pythagorean theorem to $\triangle CED$, we have $h^2 = d^2 - p^2$. Replacing h^2 in Eq. (1),

$$\begin{aligned} a^2 &= d^2 - p^2 + (m - p)^2 \\ &= d^2 - p^2 + m^2 - 2mp + p^2 \\ &= d^2 + m^2 - 2mp. \end{aligned} \tag{2}$$

Applying the Pythagorean theorem to $\triangle CEA$, we find that $b^2 = h^2 + (n + p)^2$, so

$$\begin{aligned} b^2 &= d^2 - p^2 + (n + p)^2 \\ &= d^2 - p^2 + n^2 + 2np + p^2 \\ &= d^2 + n^2 + 2np. \end{aligned} \tag{3}$$

Now multiply Eq. (2) by n to get

$$a^2n = d^2n + m^2n - 2mnp \tag{4}$$

and multiply Eq. (3) by m to get

$$b^2m = d^2m + n^2m + 2mnp. \tag{5}$$

Adding (4) and (5) gives

$$\begin{aligned} a^2n + b^2m &= d^2n + d^2m + m^2n + n^2m + 2mnp - 2mnp \\ &= d^2(n + m) + mn(m + n). \end{aligned}$$

Because $m + n = c$,

$$a^2n + b^2m = d^2c + mnc,$$

or

$$a^2n + b^2m = c(d^2 + mn).$$

To find x in $\triangle PQR$, students simply substitute into Stewart's theorem:

$$\begin{aligned} a^2n + b^2n &= c(d^2 + mn), \\ 36(4) + 81(8) &= 12(x^2 + 32), \\ 144 + 648 &= 12x^2 + 384, \\ 12x^2 &= 408, \\ x^2 &= 34, \\ x &= \sqrt{34}. \end{aligned}$$

You may wish to show students how difficult it would be to find x without Stewart's theorem. Using the figure from the foregoing proof, the students would have to find h and p before finding x.

Stewart's theorem can be used in its given form to find the length of a median of a triangle. However, because $m = n = \frac{c}{2}$, it's possible to simplify the equation, which will simplify the computation. Substituting $\frac{c}{2}$ for m and n in Stewart's theorem gives

$$\begin{aligned}
\frac{a^2c}{2} + \frac{b^2c}{2} &= c\left(d^2 + \frac{c^2}{4}\right), \\
2a^2c + 2b^2c &= 4cd^2 + c^3, \\
4cd^2 &= 2a^2c + 2b^2c - c^3, \\
d^2 &= \frac{c(2a^2 + 2b^2 - c^2)}{4c} \\
&= \frac{1}{2}a^2 + \frac{1}{2}b^2 - \frac{1}{4}c^2.
\end{aligned}$$

For $\triangle XYZ$,

$$\begin{aligned}
d^2 &= \frac{1}{2}(25) + \frac{1}{2}(81) - \frac{1}{4}(100) \\
&= \frac{25}{2} + \frac{81}{2} - 25 \\
&= 53 - 25 = 28
\end{aligned}$$

and $d = 2\sqrt{7}$.

Extension

The bisector of an angle of a triangle divides the opposite sides into segments whose measures are proportional to the measures of the other two sides of the triangle. Thus, in $\triangle ABC$,

$$\frac{a}{b} = \frac{m}{n} \quad \text{and} \quad an = bm.$$

These values are substituted in Stewart's theorem to simplify the a^2n and b^2m terms:

$$\begin{aligned}
a^2n + b^2m &= c(d^2 + mn), \\
abm + abn &= c(d^2 + mn), \\
ab(m + n) &= c(d^2 + mn), \\
abc &= c(d^2 + mn), \\
ab &= d^2 + mn, \\
d^2 &= ab - mn.
\end{aligned}$$

For the lengths given, students will find that $d^2 = 9(15) - 6(10) = 135 - 60 = 75$ and $d = 5\sqrt{3}$.

Simson's Theorem

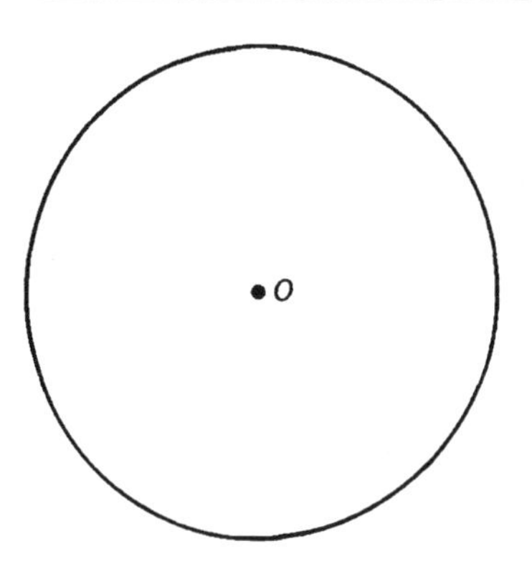

Using the preceding circle, draw any inscribed $\triangle ABC$. Choose any point P on the circle other than A, B, or C. Construct a perpendicular segment from P to each of the three sides of $\triangle ABC$. (You may have to extend the sides of the triangle.) Label the points where the perpendiculars intersect the sides X, Y, and Z. What seems to be true about points X, Y, and Z? ________________

____

The figure you have constructed illustrates Simson's theorem:

> The feet of the perpendiculars drawn from any point on the circumcircle of a given triangle to the sides of the triangle are collinear.

The line containing the three points is known as the *Simson line*. How many Simson lines does a triangle have? ________________________

Suppose point P is chosen at one of the vertices of the inscribed triangle.

How would you describe the Simson line? ____________________

____

____

Now try another construction using the circle below.

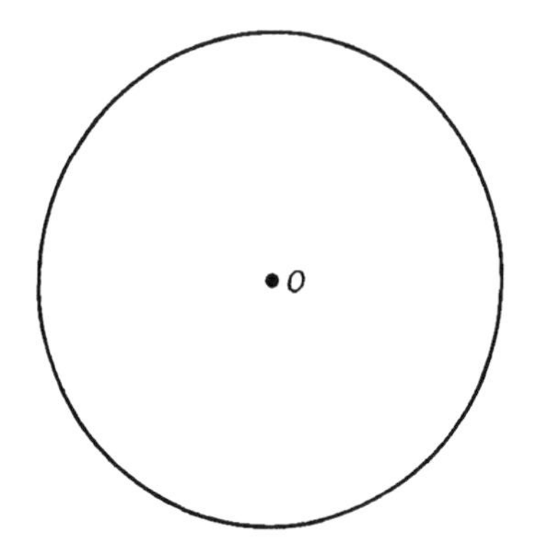

Choose any point P on circle O. Draw three chords from P meeting the circle at points A, B, and C. Construct the three circles that have $\overline{PA}$, $\overline{PB}$, and $\overline{PC}$ as diameters. These three circles intersect each other in three points. Label the points of intersection X, Y, and Z. What seems to be true about X, Y, and Z?

EXTENSION! Use Simson's theorem to prove that the three points of intersection of the three circles in the preceding problem are collinear.

Teacher's Notes for Simson's Theorem

Collinearity of points is usually a neglected topic of a high school geometry course. This is unfortunate because it's also a topic that provides many fascinating constructions and theorems. Simson's theorem is one of the more interesting theorems about collinearity of points. The activity reinforces construction techniques and angle measurement in a circle. If the proof of Simson's theorem is presented, students should review the properties of cyclic quadrilaterals.

NCTM Standards

1	2	3	4	5	6	7	8	9	10
	•	•	•		•	•		•	•

Presenting the Activity

A completed construction is

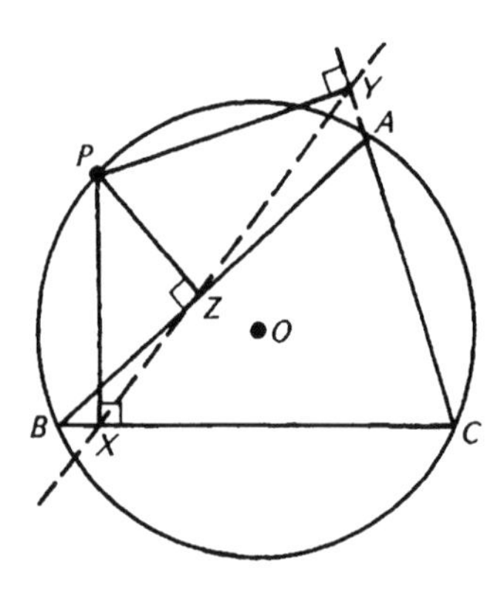

Points X, Y, and Z are collinear. Have students compare their constructions to those of other students in the class. This will emphasize that Simson's theorem is true.

Every triangle has an infinite number of Simson lines because there will be a different Simson line for every point P on the circle and there are infinitely many points P. If P is chosen at one of the vertices of the inscribed triangle, the Simson line contains the altitude from that vertex to the opposite side. Ask students which points P lie on their own Simson lines. They should see that the three vertices of the triangle are the only points for which this is true.

You may wish to present a proof of Simson's theorem. Proving collinearity of points is often difficult and the following proof uses a somewhat unusual approach that will be helpful in later work.

In the preceding figure, $\triangle ABC$ is inscribed in circle O, P is on circle O, $\overleftrightarrow{PY} \perp \overleftrightarrow{AC}$ at Y, $\overleftrightarrow{PZ} \perp \overleftrightarrow{AB}$ at Z, and $\overleftrightarrow{PX} \perp \overleftrightarrow{BC}$ at X. We want to prove points X, Y, and Z are collinear.

1. $\angle PYA$ is supplementary to $\angle PZA$ (both are right angles).
2. Quadrilateral $PZAY$ is cyclic (opposite angles are supplementary).
3. Draw $\overline{PA}$, $\overline{PB}$, and $\overline{PC}$.
4. $m\angle PYZ = m\angle PAZ$ (both are inscribed in the same arc).
5. $\angle PYC$ is supplementary to $\angle PXC$ (both are right angles).

6. Quadrilateral $PXCY$ is cyclic (opposite angles are supplementary).
7. $m\angle PYX = m\angle PCB$ (both are inscribed in the same arc).
8. $m\angle PAZ(m\angle PAB) = m\angle PCB$ (both are inscribed in the same arc).
9. $m\angle PYZ = m\angle PYX$ (transitivity with steps 4, 7, and 8).

Because both $\angle PYZ$ and $\angle PYX$ share the same ray $\overrightarrow{YP}$ and have the same measure, their other rays $\overrightarrow{YX}$ and $\overrightarrow{YZ}$ must coincide. Therefore, points X, Y, and Z are collinear.

Students may be interested to learn that Simson's theorem was not discovered by a mathematician named Simson as might be expected. It was discovered by William Wallace in 1797, but through careless misquotes was attributed to Robert Simson. Simson published an English version of Euclid's *Elements* in 1756 and his influence as a result was so great that he was mistakenly given credit for Wallace's theorem.

In the next construction, points X, Y, and Z are collinear. Again have students compare their constructions to those of other students in the class.

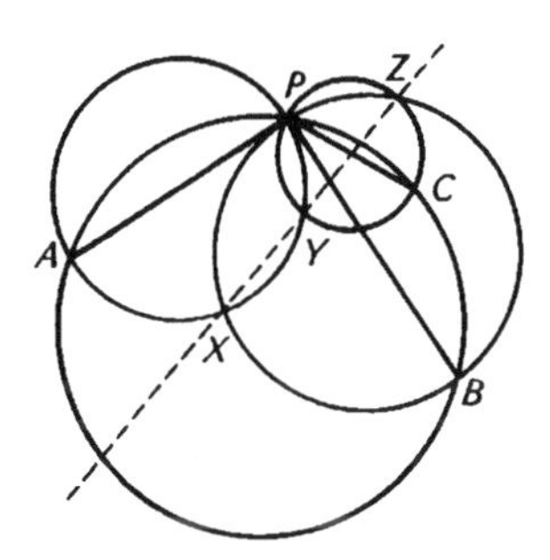

Extension

The proof is not difficult, but students may need some help deriving the relationships from the figure. Begin by drawing $\triangle ABC$ and $\overline{PX}$, $\overline{PY}$, and $\overline{PZ}$ as shown:

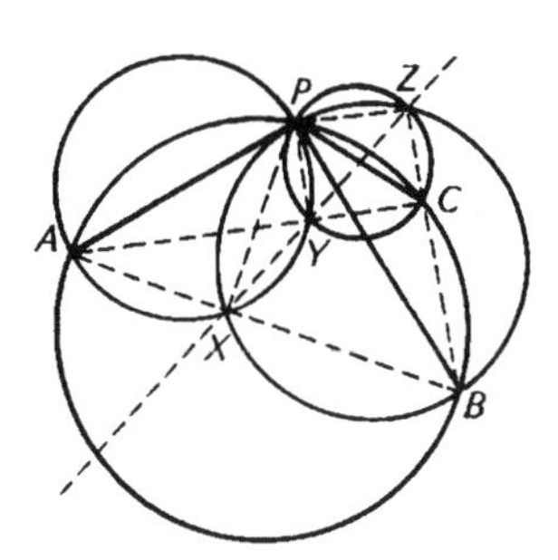

In the circle on $\overline{PA}$, $\angle PXA$ is a right angle because any angle inscribed in a semicircle is a right angle. Therefore, $\overline{PX} \perp \overline{AB}$. (*Note*: $\angle PXB$ is also a right angle, so $\overline{AXB}$ must be a straight line.) Similarly, $\angle PYC$ and $\angle PZC$ are right angles and $\overline{PY} \perp \overline{AC}$ and $\overline{PZ} \perp \overline{BC}$. (To prove Z is on $\overleftrightarrow{BC}$, note that $\angle PZC$ and $\angle PZB$ are right angles with $\overrightarrow{ZP}$ as one ray of the angle; $\overrightarrow{ZCB}$ must be the other ray.) Because $\overline{PX}$, $\overline{PY}$, and $\overline{PZ}$ are drawn from a point on the circumcircle perpendicular to the sides of $\triangle ABC$, X, Y, and Z determine a Simson line and are therefore collinear.

Napoleon's Theorem

Napoleon Bonaparte, the French general and emperor, was also interested in geometry. He is supposed to have discovered the construction shown in the following figure:

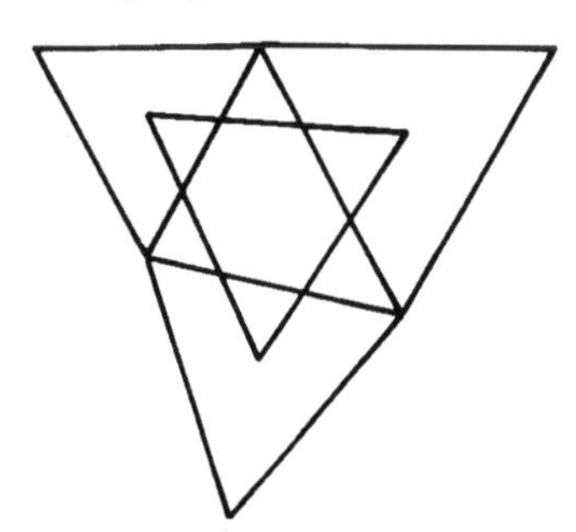

Draw any scalene triangle. On each side of the triangle construct an equilateral triangle facing outwards. Then find the center of the circumscribed circle of each equilateral triangle. Now connect the centers to form another triangle. This last triangle is also equilateral.

This construction seems rather amazing because we started with *any* scalene triangle. In addition, there are some other surprising relationships in this figure. First, consider the figure

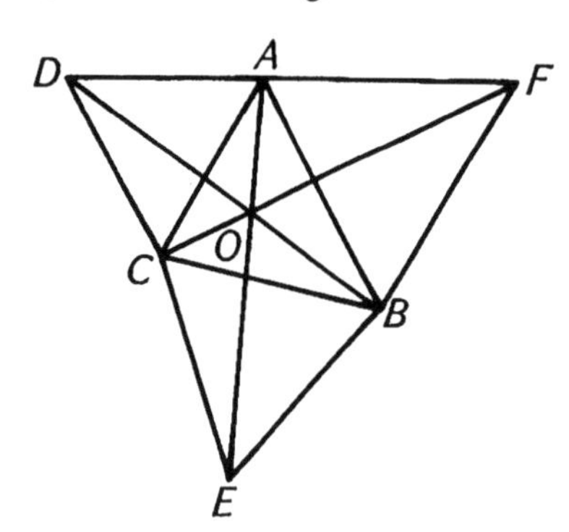

Triangle ABC is our original scalene triangle. Triangles ABF, BCE, and CAD are the equilateral triangles constructed on the sides of $\triangle ABC$. Next, $\overline{AE}$, $\overline{BD}$, and $\overline{CF}$ were drawn.

Do $\overline{AE}$, $\overline{BD}$, and $\overline{CF}$ appear to be congruent? ______________________

What triangles would you use to prove $\overline{BD} \cong \overline{CF}$? ______________________

How would you prove these triangles congruent? ______________________

____

Now prove $\overline{AE} \cong \overline{BD} \cong \overline{CF}$. ______________________________

____

Notice that $\overline{AE}$, $\overline{BD}$, and $\overline{CF}$ also appear to be concurrent (meet in a point). To prove this, we need to draw the circumscribed circle of each of the equilateral triangles, as shown:

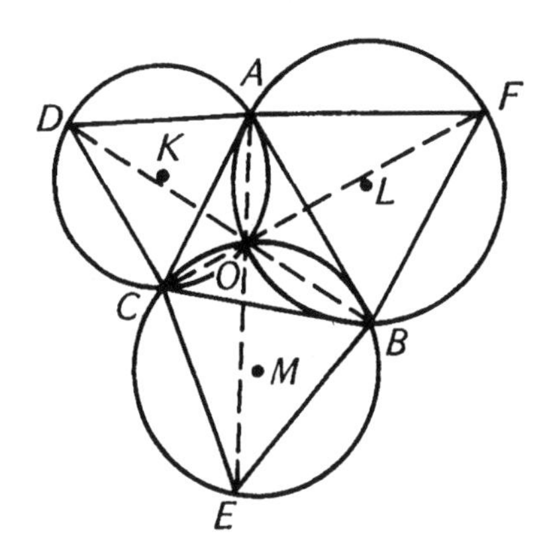

How can you find the centers K, L, and M? ______________________

To prove the segments are concurrent, we first want to show that the circles are concurrent at O. Circles K and L meet at points O and A. We need to prove that circle M also contains point O.

In quadrilateral $AOCD$, $m\angle AOC = \frac{1}{2}\widehat{ADC} =$ ______________________.

In quadrilateral $AOBF$, $m\angle AOB =$ ______________________.

Why must $m\angle BOC = 120°$? ______________________

Using arc CEB, how would you prove circle M contains point O? ________

____

Thus, the circles are concurrent at O, and $m\angle DOA =$ _______ $m\angle AOF =$ _______ $m\angle FOB =$ _______. Then $m\angle DOB =$ _______. What is true about points D, O, and B? ______________________

Is this also true for points A, O, and E and points C, O, and F? _________

Are $\overline{AE}$, $\overline{BD}$, and $\overline{CF}$ concurrent? Why or why not? ________________

____

Point O is called the *equiangular* point of $\triangle ABC$, because $m\angle AOC = m\angle BOC = m\angle AOB$.

EXTENSION! Prove that $\triangle KLM$ is equilateral.

Teacher's Notes for Napoleon's Theorem

This activity develops interesting geometric relationships from an unusual construction. Students may find these relationships difficult to believe and, therefore, should look forward to proving them. To complete these proofs, students must draw on several areas of geometric knowledge—angle measure in a circle, the basic properties of congruence and similarity, and the ratios in 30°-60°-90° triangles. For this reason, the activity should usually be presented late in a geometry course.

NCTM Standards

1	2	3	4	5	6	7	8	9	10
•	•	•	•		•	•		•	

Presenting the Activity

If possible, construct Napoleon's figure on the chalkboard before class begins or prepare a construction to use on an overhead projector. Discuss the figure in class, asking students to explain how each construction is done.

Although the next relationship uses only elementary concepts, most students find it challenging. What seems to be most perplexing is the selection of the correct pair of triangles to prove congruent. If, after a few minutes, students do not find them, tell them to name triangles that use the required segments $\overline{BD}$ and $\overline{CF}$ as sides. They will soon realize that they must prove $\triangle CAF \cong \triangle DAB$. Point out that overlapping triangles usually share a common element. Here the common element is $\angle CAB$. Whereas $\triangle ACD$ and $\triangle ABF$ are equilateral, $m\angle DAC = 60°$, $m\angle FAB = 60°$, and $m\angle DAB = m\angle FAC$ (addition). Also, $\overline{AD} \cong \overline{AC}$ and $\overline{AB} \cong \overline{AF}$. Therefore, $\triangle CAF \cong \triangle DAB$ by side–angle–side and, thus, $\overline{BD} \cong \overline{CF}$. To prove all three segments congruent, students should prove $\triangle CAE \cong \triangle CDB$ to get $\overline{AE} \cong \overline{BD}$. Then, $\overline{AE} \cong \overline{BD} \cong \overline{CF}$.

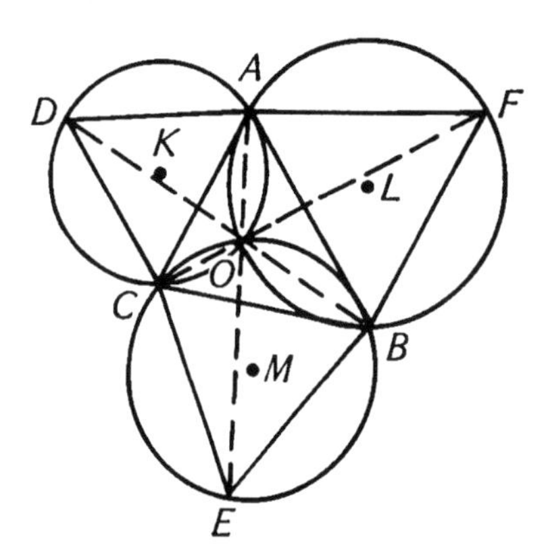

If Napoleon's figure was discussed in the beginning of the class, students should recall that the centers K, L, and M of the circumscribed circles are the point of intersection of the perpendicular bisectors of the sides of the triangles. Circles K and L meet at O and A. We need to prove circle M also contains O. Because $\triangle ADC$ is equilateral, we know $m\overset{\frown}{ADC} = \frac{2}{3} \cdot 360° = 240°$ and $m\angle AOC = 120°$. Similarly, $m\angle AOB = 120°$. Whereas a complete revolution equals 360°, $m\angle COB$ must be 120°.

In circle M, $\triangle CEB$ is equilateral. Thus, $m\widehat{CEB} = 240°$. We showed that $m\angle COB = 120° = \frac{1}{2}m\widehat{CEB}$.

Therefore, $\angle COB$ must be an inscribed angle and circle M must contain O. Therefore, we can see that the three circles are concurrent, intersecting at O.

Angles DOA, AOF, and FOB are all inscribed in arcs measuring 120°. Therefore, we know that $m\angle DOA = m\angle AOF = m\angle FOB = 60°$. This means $m\angle DOB = 180°$ and points D, O, and B are on the same line. This is also true for points A, O, and E, and points C, O, and F. Thus, $\overline{DB}$, $\overline{AE}$, and $\overline{CF}$ all contain point O and are concurrent at O.

Extension

This proof is quite difficult and it is unlikely any students will think of it on their own. First redraw the figure as follows:

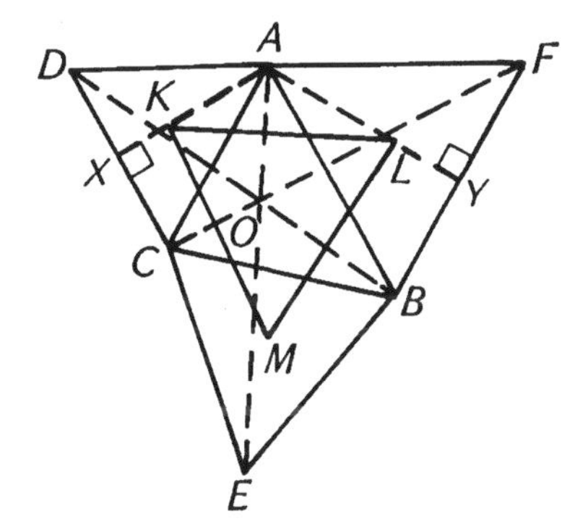

Now consider 30°-60°-90° triangle ACX:

AK is $\frac{2}{3}$ of altitude AX. Thus,

$$\frac{AC}{AK} = \frac{2}{\frac{2}{3}\sqrt{3}} = \frac{3}{\sqrt{3}} = \frac{\sqrt{3}}{1}.$$

Similarly, in $\triangle AFY$, $\frac{AF}{AL} = \frac{\sqrt{3}}{1}$. Therefore,

$$\frac{AC}{AK} = \frac{AF}{AL}.$$

We now use this to prove $\triangle KAL \sim \triangle CAF$: $m\angle KAC = m\angle LAF = 30°$ and $m\angle CAL = m\angle CAL$ (reflexive), so $m\angle KAL = m\angle CAF$ by addition. Thus, two sides of the triangles are proportional and the included angles are equal and $\triangle KAL \sim \triangle CAF$. Thus, the sides are all proportional and

$$\frac{CF}{KL} = \frac{AC}{AK} = \frac{\sqrt{3}}{1}.$$

In the same way, using other triangles in the figure, we can prove

$$\frac{DB}{KM} = \frac{\sqrt{3}}{1} \quad \text{and} \quad \frac{AE}{ML} = \frac{\sqrt{3}}{1}.$$

Therefore,

$$\frac{CF}{KL} = \frac{DB}{KM} = \frac{AE}{ML}.$$

However, because $CF = DB = AE$, as proved earlier, we obtain $KL = KM = ML$. Therefore, $\triangle KML$ is equilateral.

CHAPTER 4

Non-Euclidean Geometry

- Taxicab Geometry ♦
- Transformational Geometry—Symmetry ♦
- Projective Geometry
- Spherical Geometry

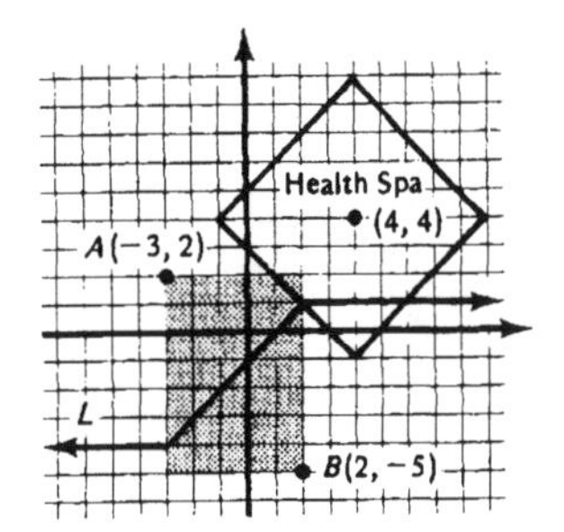

Traditional geometry courses present a somewhat one-sided view of geometry; students typically learn only the definitions and theorems of Euclidean geometry. As little as 75 years ago, non-Euclidean geometries were considered oddities and of little importance. Einstein's theory of relativity has shown the physical relevance of non-Euclidean geometries, but there has been little change in the geometry curriculum. It isn't possible to present a comprehensive study of non-Euclidean geometries, but students should at least be made aware that they exist.

The four activities in this category present four different geometries. The ideas presented are not difficult and they can be interspersed throughout the year. This gives students a change from standard Euclidean theorems and proofs, and provides them with a broader view of geometry.

"Taxicab Geometry" presents a mathematical model of a city and explores how distances are found when travel must be along city streets. Little knowledge of geometry is required for this activity, but students may need to briefly review how to locate points on coordinate axes. As mentioned in the Teacher's Notes, *Taxicab Geometry* by Eugene F. Krause is an excellent resource for students who wish to explore other ideas of taxicab geometry.

Transformational geometry is not actually non-Euclidean; it's a different approach to Euclidean geometry based on movement in a plane. "Transformational Geometry—Symmetry" presents only a very small part of transformational geometry, but it is especially applicable to the study of quadrilaterals and regular polygons. Students enjoy the change of pace from theorems about these figures, and the Extension helps to sort out the various quadrilaterals.

Although projective geometry was not fully developed until the 19th century, it originated with the Renaissance painters. Even though mathematics and painting seem to be very different fields today, in the 15th century the best mathematicians, architects, and engineers were also the best painters. The concepts developed by the Renaissance artists are still used in current books on perspective drawing. The origin of projective geometry shows students how an important branch of mathematics developed by thinking about a specific problem in the "nonmathematical" world.

The two theorems presented in "Projective Geometry" are easy to understand and the activity can be presented at any time in a geometry course. Desargues' theorem and its converse can be used to discuss duality—one of the most fascinating principles of mathematics. Morris Kline's essay in *The World of Mathematics* (J. R. Newman, ed., NY: Simon & Schuster, 1956, pp. 622–641) provides additional information on the origin, development, and importance of projective geometry.

"Spherical Geometry" gives an overview of some of the differences between spherical and Euclidean geometry. This activity more specifically compares the two geometries than do the other activities in this chapter. Thus, students must have more background in Euclidean geometry, and "Spherical Geometry" is best used only after spheres have been introduced.

Taxicab Geometry

"It's eight blocks to the subway station from my office." "The Obitz' house is four blocks from ours." "I have to walk 18 blocks to school." All these distances are the actual lengths a person has to walk from one place to another. *Taxicab geometry* is based on these distances instead of a distance that's sometimes called "as the crow flies."

The following grid represents a map of a city.

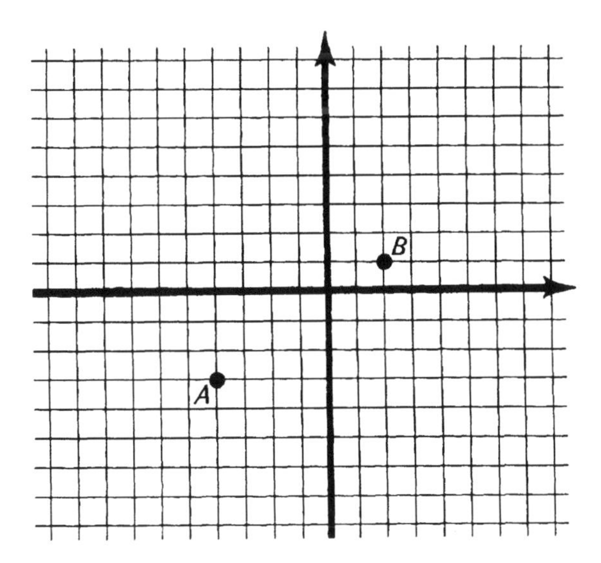

The center of town is where the axes cross. Think of the lines of the grid as streets. The coordinates of A are $(-4, -3)$. What are the coordinates of B? _______ Mark P at $(-2, -1)$. What is the taxi distance from A to P? _______ Mark all the points you can find at a taxi distance 4 from A. Connect these points to form a square. This figure is a taxi *circle*.

What is the shortest taxi distance from A to B? _______ Mark C at $(-3, 1)$. What is the sum of the taxi distance from C to A plus the taxi distance from C to B? _______ Mark all the points you can find so that the sum of the taxi distance from the point to A plus the point to B is 10. Where are all these points located?

____

Now mark all the points that are an equal taxi distance from A and B.

Use a separate sheet of graph paper to solve this problem: Amanda and Brian are looking for an apartment. Amanda works at a store at $(-3, 2)$. Brian works in an office at $(2, -5)$. They would like their apartment to be located so that the distance Amanda has to walk to work plus the distance Brian has to walk to work is as small as possible. Show where their apartment could be located on your graph.

After a day of apartment hunting, they decide all they really need is for them both to be the same distance from their jobs. Where could their apartment be located?

____

They find an apartment that is five blocks from their health spa and equidistant from their jobs. The health spa is at $(4, 4)$. Where is the apartment? ________

____

EXTENSION! There are three voting locations in a city, at $(-6, 4)$, $(4, 2)$, and $(-3, -9)$. On a separate sheet of graph paper draw the precinct boundary lines so that each person can vote at the location closest to his or her home. After the polls close, the ballots are taken to a central location for counting. What point is an equal distance from each voting location? The population in the city increases and two new voting locations are opened at $(7, 7)$ and $(6, -6)$. Redraw the precinct boundary lines.

Teacher's Notes for Taxicab Geometry

Euclidean geometry is the geometry of a perfect world and most students see no relationship between geometry and the imperfect world we live in. Most non-Euclidean geometries try to relate geometry to our imperfect world. However, most non-Euclidean geometries are difficult for secondary school students to understand. Taxicab geometry is an exception. Students will be delighted to find a geometry that is based on their familiar world and that has square circles instead of round ones.

This activity explores only one idea of taxicab geometry—the distance between two points. Other geometric figures and ideas can also be considered. An excellent resource book for further exploration is ***Taxicab Geometry*** *by Eugene F. Krause (Addison-Wesley, Reading, MA, 1975).*

NCTM Standards

1	2	3	4	5	6	7	8	9	10
•	•	•	•		•			•	•

Presenting the Activity

Briefly review how points are located on coordinate axes. Then have students begin the student page. Discuss how the grid differs from a real city: On the grid, buildings have no size, streets have no width, all the streets run exactly north and south or east and west, and all the blocks are the same size. Although the grid is only an ideal model of a city, many of the problems solved using taxicab geometry have applications in real situations. For example, when an automobile accident occurs, the police dispatcher must know which police cars are closest to the scene of the accident.

The coordinates of B are (2, 1). The taxi distance from A to P is 4. Have students find some other taxi distances. They should quickly grasp the idea of counting blocks to find taxi distances. The taxi circle with center A and radius 4 is

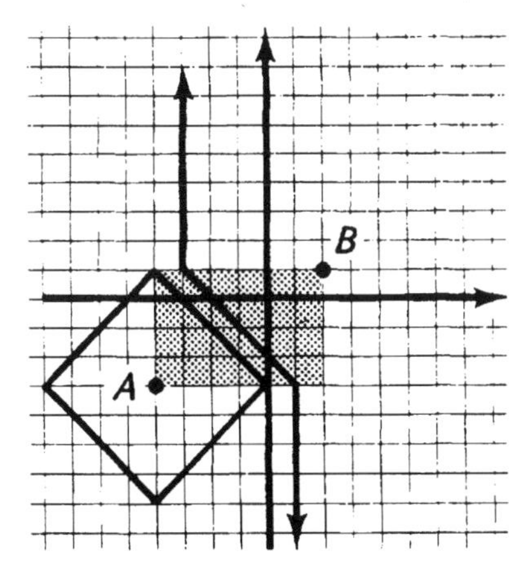

Notice that many of the points on a taxi circle are not on a street. For example, $(-3\frac{1}{2}, \frac{1}{2})$ is in the middle of a block. Nevertheless, the taxi distance to this point from A (assuming a taxi could drive to it), would be 4. The taxi distance between points (x_1, y_1) and (x_2, y_2) is defined algebraically as $|x_1 - x_2| + |y_1 - y_2|$. When a problem asks for *all* the points at a given taxi distance from a point, the points not on streets should be included.

The shortest taxi distance from A to B is 10. This is also the sum of the taxi distances from C to A plus C to B. The shaded rectangle on the preceding grid includes all the points such that the sum of the taxi distances from the point to A plus the point to B is 10. The line on the grid is all the points an equal taxi distance from A and B.

The completed graph of the next problem is

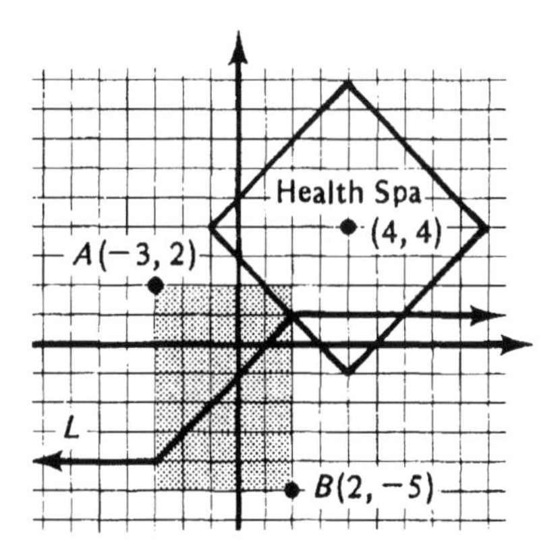

The shaded rectangle is the area of Amanda's and Brian's first search for an apartment. The points equidistant from their jobs are on L. The points five blocks from the health spa are on the taxi circle with center (4, 4) and radius 5. Thus, their apartment is at either (6, 1) or (2, 1).

Extension

Students first locate the three points on their graphs. Then they find the points equidistant from two of the points, then the points equidistant from a different pair of points, and finally the points equidistant from the last pair of points. By testing points, they can find which line to use for regions that overlap. The precinct boundary lines for the three voting locations are

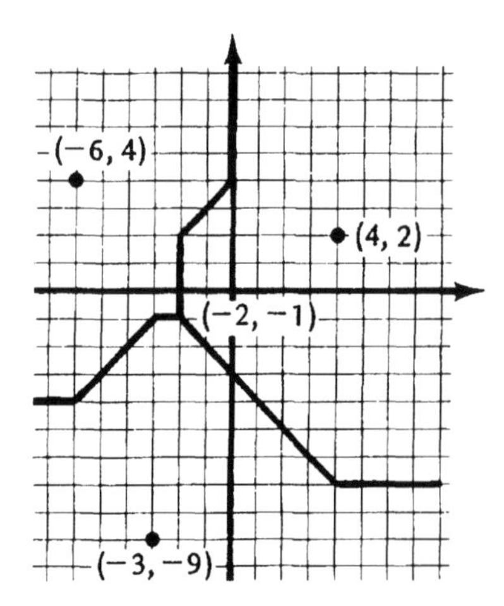

The central location for counting the votes is at (−2, −1).

When the two new voting locations are added, the precinct boundary lines are

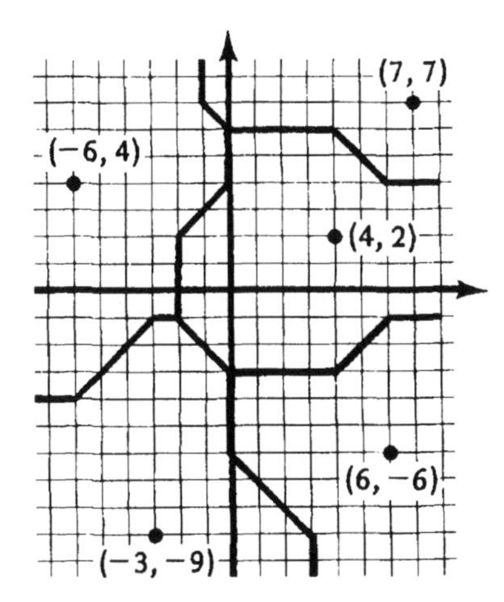

The previous regions are altered by finding the points equidistant from each new point and the two points closest to each. Note that there is no point equidistant from the five locations.

Transformational Geometry—Symmetry

If you drew a vertical line through the center of a photograph of your face, would the two halves of your face match exactly? If they did, your face would have *reflection* or *line symmetry*. Your face probably doesn't have line symmetry, but many geometric figures do have lines of symmetry. An isosceles triangle such as

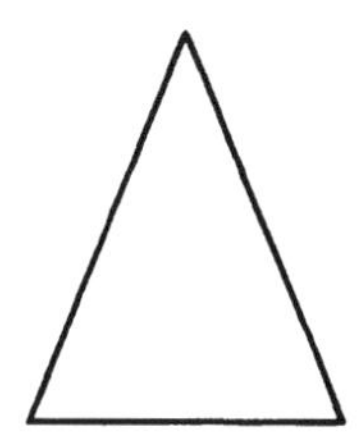

has only one line of symmetry. Draw this line and describe where it lies.

____

Some figures have a different kind of symmetry called *rotation symmetry*. In the following diagram, if the figure at the left were rotated 90° about point A, it would coincide exactly with the original figure;

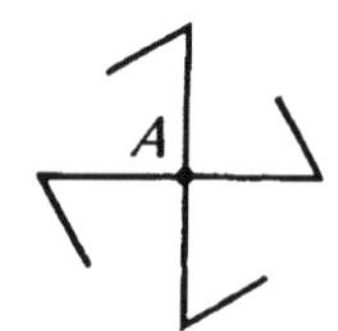

that is, it would look just the same. The center of rotation is A and the degree of rotation is 90°. An equilateral triangle has rotation symmetry. The center of rotation is the center of the triangle's circumscribed circle. What is the degree of rotation? _______

The capital letters of our alphabet are shown below.

ABCDEFGHIJKLMNOPQRSTUVWXYZ

Which letters have a vertical line of symmetry? ____________________

Which have a horizontal line of symmetry? ______________________

Which have *both* horizontal and vertical lines of symmetry? ____________

Do any of the letters have rotation symmetry? If so, which ones? _________

Write a *word* that has a vertical line of symmetry: __________________

Write a word that has a horizontal line of symmetry: ____________________

Write a word that has rotation symmetry: ____________________________

Regular polygons have both reflection and rotation symmetry. Draw all the lines of symmetry for each of the following regular polygons.

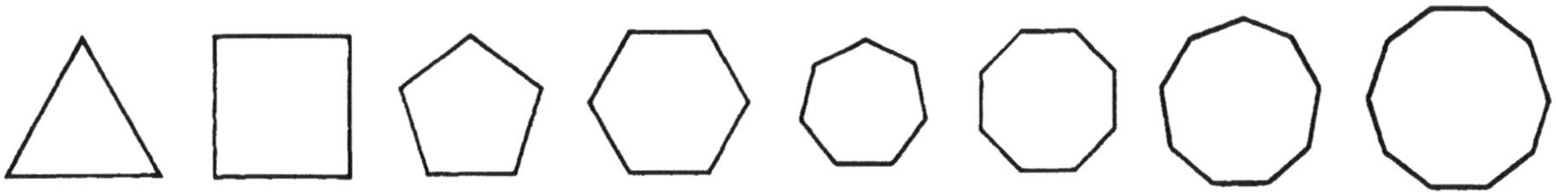

How does the number of lines of symmetry compare to the number of sides of the polygon? ____

Find the *degree* of rotation for the square, the pentagon, and the hexagon: ____

____

Write an equation for the degree of rotation for a regular polygon in terms of n, the number of its sides: ________________________________

EXTENSION! A square is a regular quadrilateral. It has four lines of symmetry and has rotation symmetry. Draw a quadrilateral with (1) no lines of symmetry and no rotation symmetry, (2) rotation symmetry and no lines of symmetry, (3) one line of symmetry and no rotation symmetry, (4) rotation symmetry and two lines of symmetry.

Teacher's Notes for Transformation Geometry—Symmetry

Symmetry is only a small part of transformational geometry. However, it presents ideas that many students will find very interesting. This activity provides a good introduction to transformational geometry and can be expanded to show students how Euclidean geometry can be approached from a different viewpoint. You can also approach the activity as a change of pace from standard proofs and theorems.

NCTM Standards

1	2	3	4	5	6	7	8	9	10
		•					•		

Presenting the Activity

A person's face is very close to symmetric, but the left and right sides will not match exactly. The nose may be a little crooked or one eyebrow will be higher or differently shaped. If possible, bring a front view photo of a person to class with a vertical line drawn through its center. Ask students to find points of dissimilarity between the two sides.

The line of symmetry for an isosceles triangle is the line through the vertex and the midpoint of the base. It can easily be proved that the two triangles formed are congruent and thus, one will exactly match the other. Using an overhead projector is an excellent way to present rotation symmetry. Prepare two congruent equilateral triangles on separate transparencies, marking the center of rotation on each and drawing a line from the center of rotation to one vertex. When the triangle is rotated to coincide with its image, the lines will have formed an angle of 120°.

Most students should be able to find the symmetries in the letters of the alphabet with little difficulty. The letters with vertical lines of symmetry are A, H, I, M, O, T, U, V, W, X, and Y. Those with horizontal lines of symmetry are B, C, D, E, H, I, K, O, and X. Those with both horizontal and vertical lines of symmetry are of course the ones that occur in both lists: H, I, O, and X. These four letters also have rotation symmetry. The other letters with rotation symmetry are N, S, and Z.

Several examples of words with line and rotation symmetry are

MOM WOW AHA TOOT

BOX HIDE CHOICE

NO•ON SIS ONO

Ask students what rules must be used to form the words: For words with a horizontal line of symmetry, any combination of letters with a horizontal line of symmetry may be used. However, for words to have rotation symmetry or a vertical line of symmetry, the words must read the same backward and forward in addition to the letters coming from the correct lists.

The polygons and their lines of symmetry are

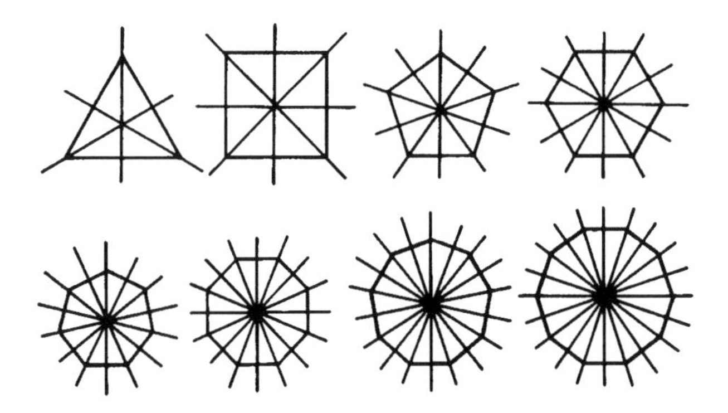

The number of lines of symmetry equals the number of sides (or vertices) of the polygon.

The degree of rotation for the square is 90°; for the pentagon, 72°; and for the hexagon, 60°. Thus, the degree of rotation for a regular polygon of n sides is $\frac{360°}{n}$. Notice that we have used the degree of rotation to mean the *smallest* rotation necessary for the figure to coincide with its image. Each figure has other degrees of rotation. For example, the square has 90°, 180°, and 270° of rotation. (Only rotations less than 360° are considered.)

Extension

Most students should be able to solve the Extension. If some students have difficulty, have them draw various quadrilaterals and check each for rotation symmetry and lines of symmetry. There is more than one answer in some of the four cases. An example for each case follows:

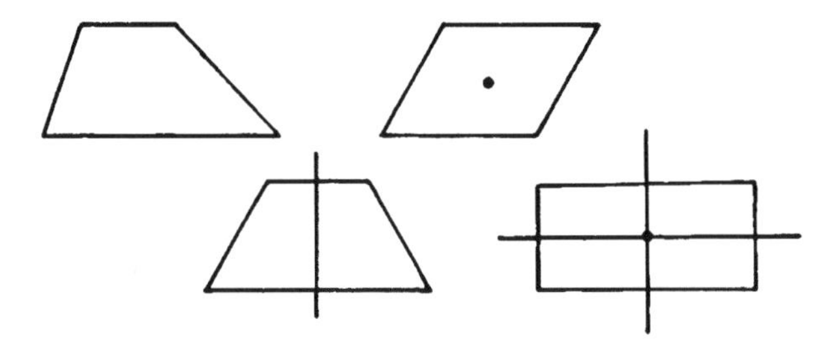

There are many alternate extensions for this activity, and some students may be interested in exploring them. For example, a third type of symmetry is *translation* symmetry. In this case, the image slides to the right or left to coincide with the figure. An electrocardiogram and a border print show translation symmetry.

Students interested in art may wish to find examples of reflection, rotation, and translation symmetry in painting and sculpture. There are many such examples.

Finally, you may wish to discuss planes of symmetry for space figures. For example, a rectangular box has three planes of symmetry. Students can investigate the planes of symmetry for various prisms and pyramids.

Projective Geometry

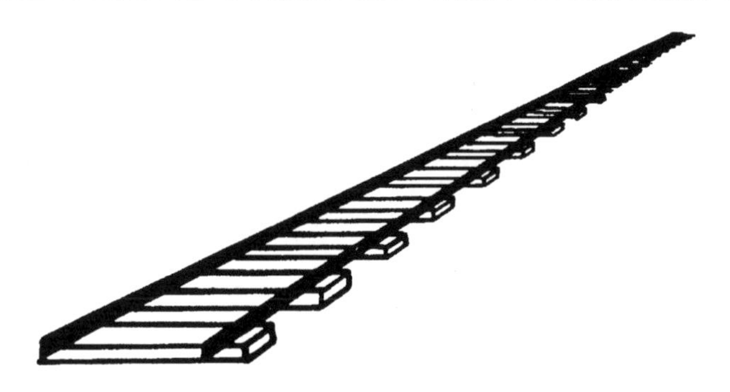

Railroad tracks are a physical example of parallel lines—you know they can never meet. The preceding figure shows railroad tracks going off into the distance. What *appears* to be true? ______________________________

____

The problem of how to represent the three-dimensional real world in a two-dimensional drawing was solved by the Renaissance painters. They imagined a glass screen between the artist and the object to be painted. The lines from the artist's eye to the object are called *lines of projection*. The set of points where the lines of projection intersect the imaginary glass screen is a *section*. The section is then reproduced on the artist's canvas.

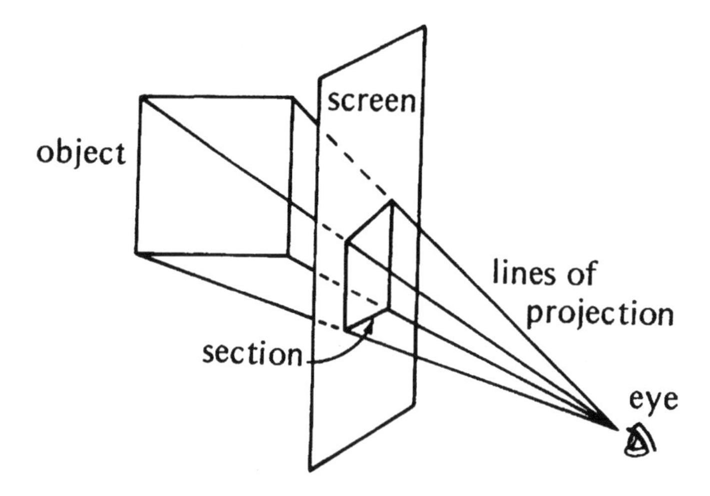

The relationships between an object and its section provided mathematicians with the basis for *projective geometry*. In projective geometry, *any* two lines meet in one and only one point. Thus, no lines are parallel. One of the basic theorems of projective geometry is Girard Desargues' theorem.

This 16th-century drawing by Albrecht Dürer illustrates one way in which the issue of perspectivity was dealt with in the 16th century.

Consider the following figure:

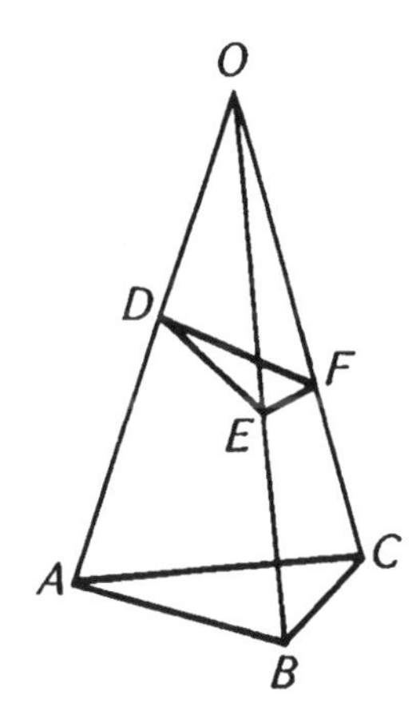

The eye at point O looks at $\triangle ABC$. $\overleftrightarrow{OA}$, $\overleftrightarrow{OB}$, and $\overleftrightarrow{OC}$ are lines of projection and $\triangle DEF$ is a section. Extend $\overline{AB}$ and $\overline{DE}$ to meet at R, $\overline{BC}$ and $\overline{EF}$ to meet at S, and $\overline{AC}$ and $\overline{DF}$ to meet at T. What is true about R, S, and T?

____

In the last figure, we assumed $\triangle ABC$ and $\triangle DEF$ were in different planes. However, Desargues' theorem is true if the triangles are both in the same plane:

> If the lines joining corresponding vertices of two triangles are concurrent, then the points of intersection of the corresponding sides are collinear.

On a separate sheet of paper, draw any two triangles "line up" (lines joining corresponding vertices are concurrent). Do the corresponding sides meet in three collinear points? ________________________

Write the *converse* of Desargues' theorem: ________________________

____

____

EXTENSION! Another theorem of projective geometry was proved by Blaise Pascal:

> If the opposite sides of any hexagon inscribed in a circle are extended, the three points at which the extended pairs of lines meet will be collinear.

Draw a figure to illustrate this theorem. (Avoid having any opposite sides parallel.)

Teacher's Notes for Projective Geometry

Projective geometry has been called "all geometry" because Euclidean geometry and the non-Euclidean geometries of Lobachevsky, Bolyai, and Riemann can all be derived as special cases of projective geometry. In addition, the basic theorems of projective geometry are easy to understand. In spite of this, most high school students never study projective geometry or learn anything of its origin.

Space limitations prevent a comprehensive study of projective geometry. It is hoped, however, that this activity will encourage students to explore this elegant and "logically perfect" geometry.

NCTM Standards

1	2	3	4	5	6	7	8	9	10
		•				•		•	•

Presenting the Activity

Most students will have had some drawing experience and most will have been frustrated trying to make their drawings reflect the real world. What we see and what we know to be true are often contradictory, as in the case of the railroad tracks.

Different sections of the same object look very different depending on the position of the viewer and the position of the glass screen. Mathematicians tried to find geometric properties common to an object and its various sections. This is the basis of projective geometry.

Ask students which properties of an object change from section to section (length, angle, area, and parallelism) and which do not (a straight line is still straight, a triangle is still a triangle). Desargues' theorem states a more significant property common to an object and its sections. When students extend the corresponding sides of the two triangles, they will find that points R, S, and T are collinear.

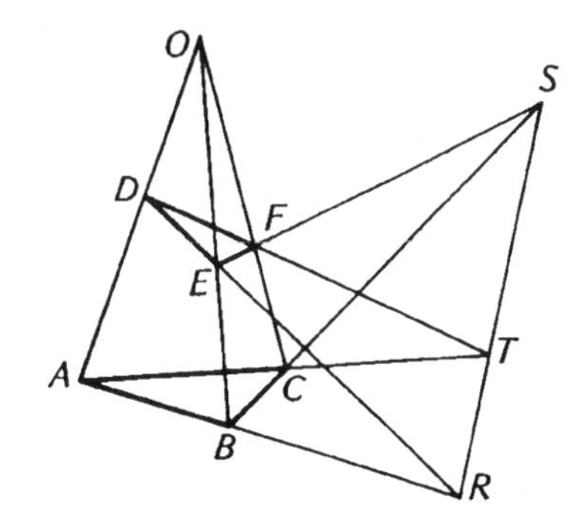

Some students may wonder what happens if the sides of the triangles happen to be parallel. (This situation may arise by accident when students draw their own "lined-up" triangles.) Explain that in projective geometry, lines are defined as meeting in a point. Lines that appear parallel should be thought of as meeting in a point at infinity. It is also agreed that all the intersection points of different sets of parallel lines lie on a "line at infinity." So even if each of the three pairs of corresponding sides of the triangles are parallel, their intersections would still be collinear on the line at infinity. These arguments are logical when students are reminded that projective geometry arose from depicting the real world as it's seen by the human eye. As shown by the railroad-tracks example, we never actually see parallel lines.

The converse of Desargues' theorem is as follows:

> If the points of intersection of the corresponding sides of two triangles are collinear, then the lines joining corresponding vertices are concurrent.

Desargues' theorem and its converse are examples of the principle of *duality*. One statement is obtained from the other by interchanging certain words:

$$\begin{aligned} \text{lines} &\longleftrightarrow \text{points} \\ \text{sides} &\longleftrightarrow \text{vertices,} \\ \text{concurrent} &\longleftrightarrow \text{collinear.} \end{aligned}$$

These words are *duals* of each other. Point out that essentially only the words, point and line, are interchanged: Sides are lines and vertices are points; concurrent means lines meeting in a point and collinear means points meeting in a line. Hence, by interchanging point and line, we have written a new theorem. The proof of the converse of Desargues' theorem is written directly from the proof of Desargues' theorem simply by interchanging point and line in each step of the proof. The principle of duality is true for *every* theorem of projective geometry.

Extension

The following figure shows more than what is asked for on the student page. This figure also shows the lines of projection meeting at a point and a section, so the relationship of Pascal's theorem to projective geometry is clear.

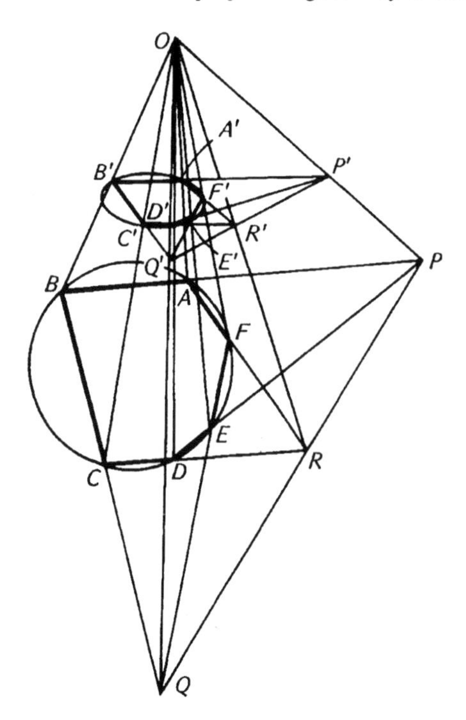

In the section, the pairs of opposite sides meet at points P', Q', and R' and P', Q', and R' are collinear. Also notice that O, P, and P' are collinear; O, Q, and Q' are collinear; and O, R, and R' are collinear.

Spherical Geometry

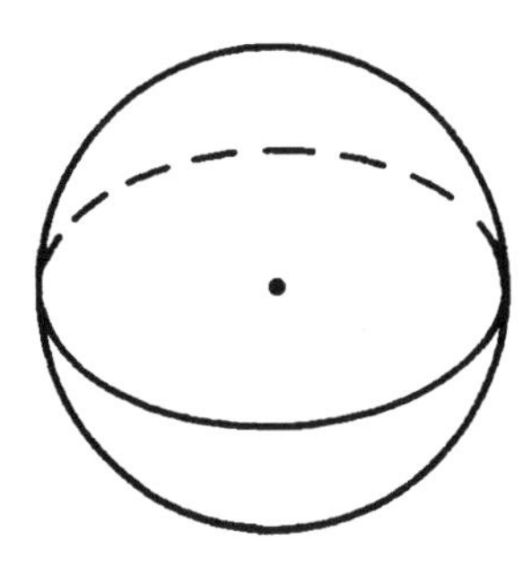

True or false:

- Through a point not on a line there is exactly one line perpendicular to the line.
- If two lines intersect, their intersection contains only one point.
- The sum of the angle measures of a triangle is 180°.

These statements are all true in the geometry developed by Euclid. They are all false in *spherical geometry*. In spherical geometry, a flat surface is not used, but rather the surface of a sphere. A line in spherical geometry is a great circle of a sphere. What is a great circle? ______________________________

____

Do spherical lines have length? __________ What is the length of a line in spherical geometry? ______________________________

A point in spherical geometry is any point on the surface of a sphere. How many points of intersection are there for two spherical lines? ______ In the figure below

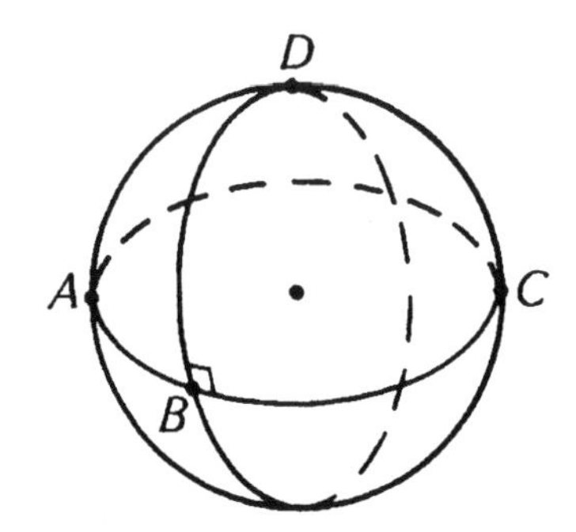

points A, B, and C are on the same great circle. Are A, B, and C collinear? __________ Which of the three points is between the other two? ______

Suppose the spherical line through D is perpendicular to line AB. Are there other lines through D perpendicular to line AB? __________ If so, how many? ____________ Are there any lines parallel to line AB? ____________

Now consider spherical triangles. In spherical triangle ABC,

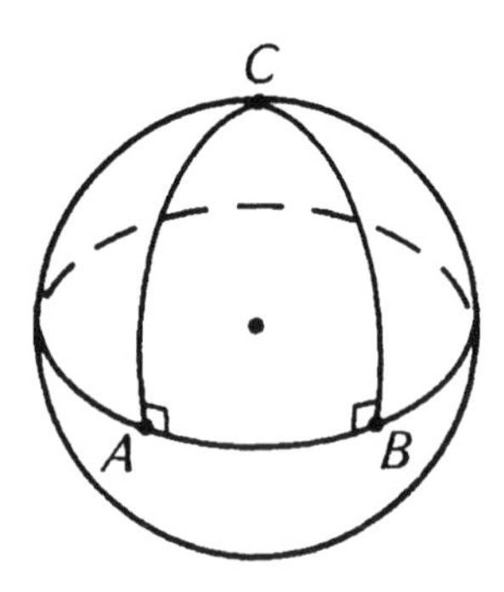

is the sum of the angle measures 180°? ____________ Is it greater or less than 180°? ____________ How many right angles can a triangle have in Euclidean geometry? ____________ How many right angles can a triangle have in spherical geometry? ____________

A triangle in Euclidean geometry can have only one obtuse angle. How many obtuse angles can a spherical triangle have? ____________ An obtuse angle has a measure greater than 90° and less than 180°. Thus, the sum of the angle measures of a spherical triangle is less than ____________.

EXTENSION! In spherical geometry, there are no parallelograms because there are no parallel lines. Suppose a rectangle is not defined as a parallelogram, but rather as an equiangular quadrilateral. Then can a rectangle exist in spherical geometry? If so, what kind of angles will it have? What other quadrilaterals from Euclidean geometry exist in spherical geometry? What quadrilaterals do not? Draw figures to illustrate your answers.

Teacher's Notes for Spherical Geometry

Spherical geometry is an example of Riemannian geometry, the geometry used by Einstein in his theory of relativity. A sphere is an easy model to work with and of course represents the world we live on. Students will be surprised to find the number of Euclidean concepts that do not apply when geometry is modeled on this universe.

"Spherical Geometry" should be presented after students have had some experience studying properties of spheres in Euclidean geometry.

NCTM Standards

1	2	3	4	5	6	7	8	9	10
		•			•	•	•		

Presenting the Activity

Have a globe and several other spheres in class to use as models. By stretching rubber bands around the spheres to represent great circles, students can more easily see the figures being considered.

A great circle of a sphere is the intersection of the sphere and a Euclidean plane containing the center of the sphere. A great circle is the shortest path between two points on a sphere. Airplane routes are a good example of this: On a flat map the routes seem longer, but following the route on a globe shows the airplanes take the shortest path. Although a spherical line has no endpoints, it does have length, the circumference of the sphere, $2\pi r$.

There are two points of intersection for two spherical lines. Collinear points in spherical geometry are the same as collinear points in Euclidean geometry. Betweenness of points is a different matter. Students may say that point B is between points A and C in the figure on the student page. However, it is possible to go from A to C without crossing point B, so it is not possible to tell which of the three points is between the other two.

Spherical geometry is sometimes presented using *polar points*. Polar points are the points of intersection of a line through the center of the sphere with the sphere. On a globe, the north and south poles are polar points. Polar points can be considered to count as just one point. When this is done, two spherical lines will have only one point of intersection as in Euclidean geometry. Also, for a given pair of points, there is exactly one line that contains them. If polar points are considered as two separate points, two points do not necessarily determine a line. In the following figures, points A and B determine a line, but there are infinitely many lines through polar points C and D.

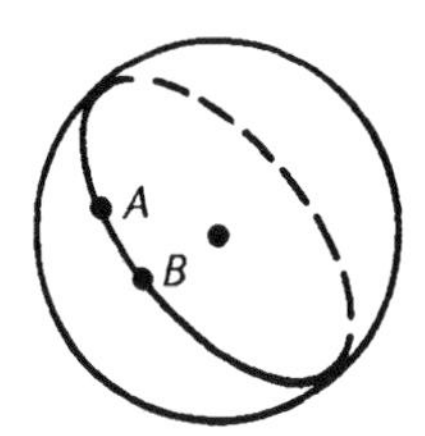

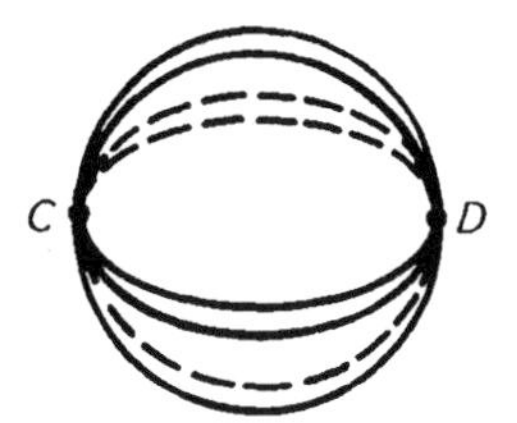

There are an infinite number of lines through D perpendicular to line AB. Ask students when there will be only one perpendicular to line AB from a point not on the line.

If D is not a polar point, but rather another point on line DB, there will be only one perpendicular. Also, through point B on line AB there is only one perpendicular as in Euclidean geometry.

In spherical geometry there are no parallel lines because two spherical lines always intersect. If students question this, remind them that a spherical line is defined as a great circle and that two great circles always intersect.

In a spherical triangle, the sum of the angle measures is always greater than 180°. Although there can be only one right angle in a Euclidean triangle, a spherical triangle can have three right angles. If students have difficulty seeing this, point out that lines AC and BC in the figure on the student page could be perpendicular at C. Similarly, a spherical triangle can have three obtuse angles and the sum of the angle measures of a spherical triangle is less than 540°. Use a globe and other actual spheres to consider the angles of various spherical triangles. Spherical triangles can have one, two, or three obtuse angles; they can have one, two, or three right angles; and they can have one, two, or three acute angles. Note that if there are three acute angles, the sum of the measures of the angles is still greater than 180°.

Ask students for some other Euclidean theorems that are not true in spherical geometry. In spherical geometry, two lines perpendicular to the same line are not parallel. Also, the measure of an exterior angle of a spherical triangle does not equal the sum of the measures of the two remote interior angles.

Extension

Have students use globes and spheres to explore spherical quadrilaterals. If a rectangle is defined as an equiangular quadrilateral, a rectangle exists in spherical geometry. It will have four obtuse angles. A spherical square will be a spherical rectangle with equilateral sides. A spherical rhombus is an equilateral quadrilateral. A trapezoid does not exist in spherical geometry, but a quadrilateral similar to an isosceles trapezoid could be defined as a quadrilateral with two pairs of equal consecutive angles with no angle common to both pairs. Of course, many Euclidean theorems about quadrilaterals will not apply to spherical quadrilaterals. The figures below illustrate the spherical quadrilaterals discussed:

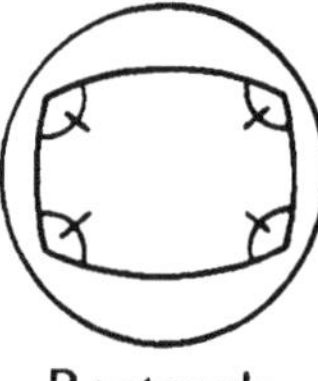
Rectangle

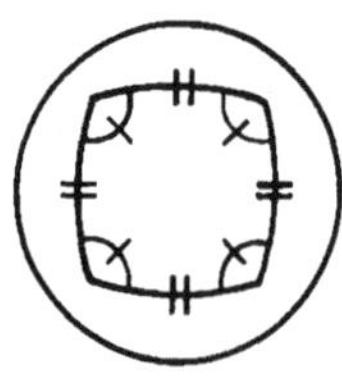
Square

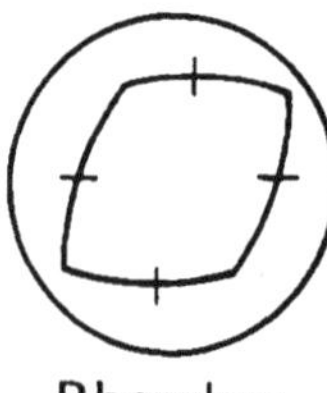
Rhombus

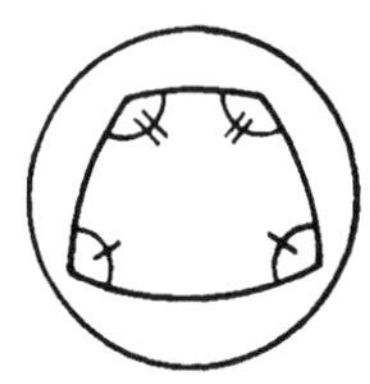
Trapezoid

CHAPTER 5

Solid Geometry

- This Wraps It Up ♦
- Regular Polyhedra
- Cavalieri's Principle
- The Jolly Green Giant? ♦

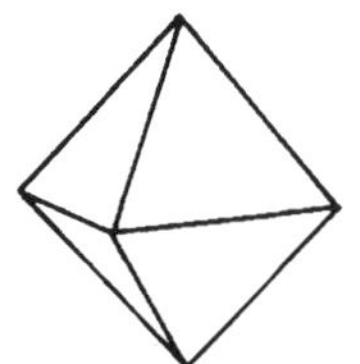
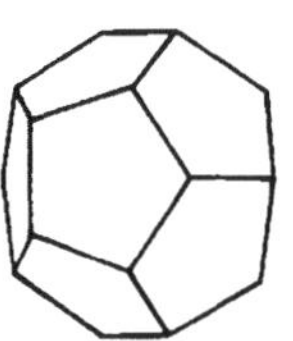

Over the past few decades or so, solid geometry has been given a progressively smaller role, and many interesting relationships and ideas simply have been dropped from the curriculum. Most texts treat little other than the surface area and volume formulas of solids. There is often no justification of these formulas and few applications. The activities in this section are not a mini course in solid geometry, but they can introduce students to some of the concepts in this branch of mathematics.

"This Wraps It Up" poses a rather deceptive problem: Which way of wrapping a package uses less ribbon? Surprisingly, finding the answer to this question doesn't involve volume or surface area, but only the Pythagorean theorem. However, because it *is* necessary for students to visualize how a rectangular box can be unfolded to produce a flat surface, the activity is a good introduction to volume and surface area. To solve the Extension, students must set up general equations using their solutions to the specific problem analyzed earlier. To solve one of these equations, they must use the quadratic formula, so the activity provides a good algebra review.

"Regular Polyhedra" leads students through an intuitive "proof," based on manipulation of models, that only five regular polyhedra can exist. This is followed by a proof based on mathematical facts. Although the former "proof" is not a proof in any formal sense, practicing scientists know that both processes are needed for scientific inquiry. Euler's theorem that relates the number of vertices, faces, and edges is also developed. A proof that only five regular polyhedra exist using Euler's theorem is given in *The World of Mathematics* (J. R. Newman, ed., NY: Simon & Schuster, 1956, pp. 584–585). "Regular Polyhedra" is very motivational and be presented as soon as regular polygons have been studied.

"Cavalieri's Principle" presents an easy-to-understand derivation of the formula for the volume of a sphere. Although some texts derive this formula using the method of this activity, they simply show students the derivation. In "Cavalieri's Principle," students must think through each step—a much more effective way to make sure students remember and understand this formula. The activity can easily be used in place of your text's presentation of the formula for the volume of a sphere.

Most textbook applications that relate surface area and volume deal with water tanks, storage bins, and other objects that are of little interest to students. Finding out whether or not giants can exist is a much intriguing problem. "The Jolly Green Giant?" explores several biological applications of the ratio of surface area to volume and will help emphasize the importance of mathematics to other scientific fields. The Extension may be difficult for some students and should be discussed thoroughly during class.

This Wraps It Up

Look at the following two ways to wrap a gift. Which do you think uses less ribbon? ______________________________

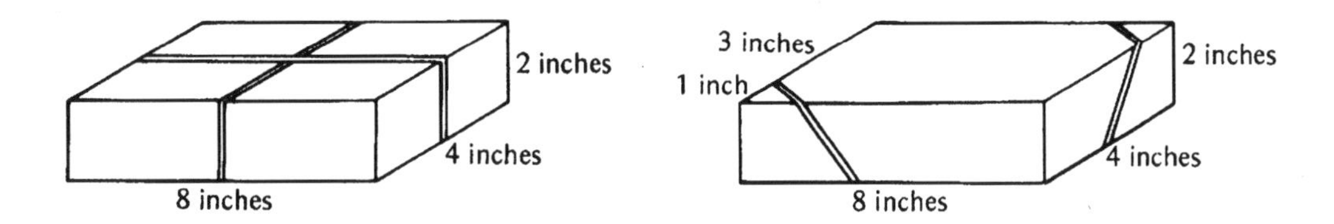

First consider the package on the left. Complete the following to find out how much ribbon is needed (not including the knot and bow).

Number of 2-in. lengths: ______________________________

Number of 4-in. lengths: ______________________________

Number of 8-in. lengths: ______________________________

Total length of ribbon: ______________________________

Finding out how much ribbon is used for the other package is a little more difficult. First a line is drawn on the box where the ribbon touches it:

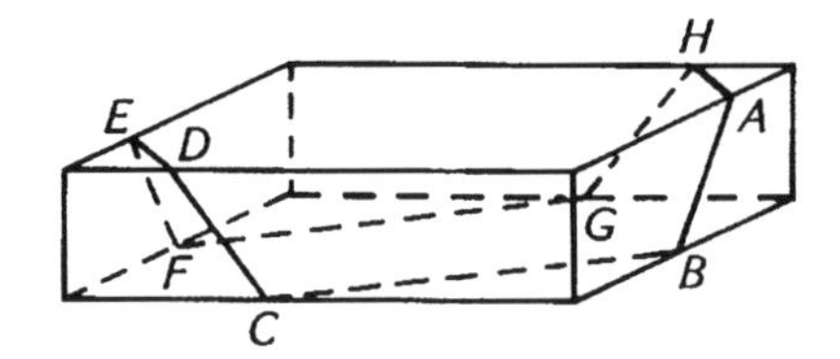

Then the box is unfolded as shown:

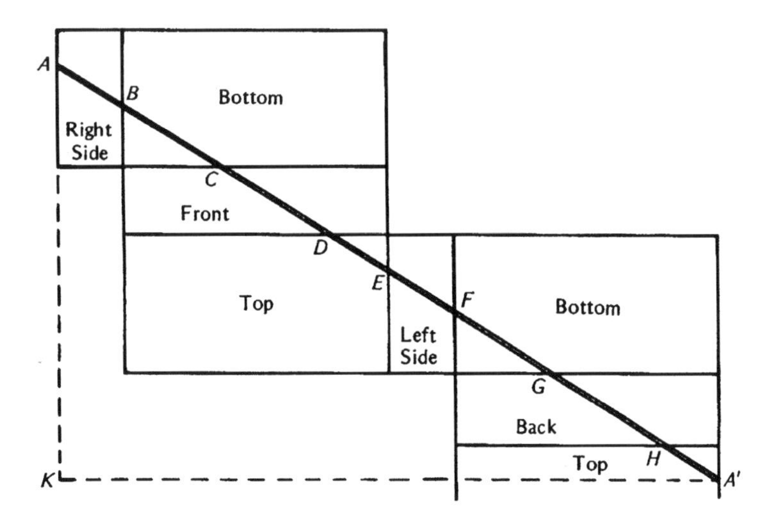

The ribbon is the line segment $\overline{ABCDEFGHA'}$. What is the length of $\overline{AK}$? ____________ What is the length of $\overline{A'K}$? ____________ What is the length of $\overline{AA'}$ to the nearest tenth of an inch? ______________________________
Which way of wrapping uses less ribbon? ____________ How much less? ______________

EXTENSION! A piece of ribbon 44 in. long is to be used to wrap a rectangular box 2 in. deep and 10 in. long. A knot and bow are not included and all the ribbon is to be used. How wide can the box be using the first method? How wide can the box be using the second method?

Teacher's Notes for This Wraps It Up

This activity is an interesting and enjoyable problem-solving challenge, and yet it should be within the reach of all your students. The only mathematical concepts needed for the main body of the activity are addition and the Pythagorean theorem. Thus, it's a good motivational or change-of-pace item. The latter half of the Extension will appeal to those of you who wish to insert some algebra review; it requires students to apply the quadratic formula.

NCTM Standards

1	2	3	4	5	6	7	8	9	10
•	•	•	•		•	•		•	•

Presenting the Activity

Most students will not find the first half of this activity at all difficult. There are four 2-in. lengths of ribbon, two 4-in. lengths, and two 8-in. lengths. Thus, the total length of ribbon is 32 in.

It will be helpful to prepare a box before class to demonstrate the second method on the student page. Cut a piece of heavy paper using the dimensions shown below for the flattened box.

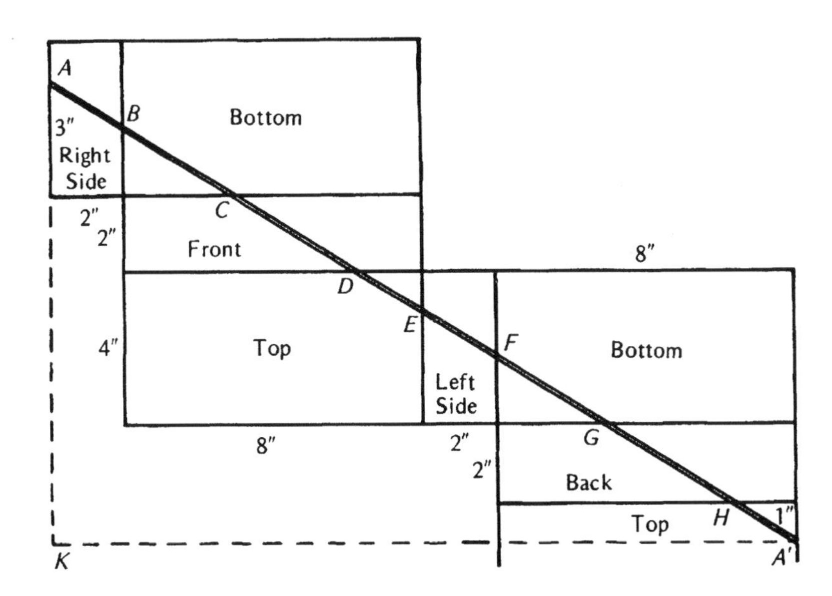

Draw and label segment $ABCDEFGHA'$ and fold up the box. Present the folded-up box to the class and then unfold it to show the ribbon line.

Have students label all the dimensions on the unfolded box as shown. The lengths of $\overline{AK}$ and $\overline{A'K}$ are found directly from the figure: $AK = 12$ in. and $A'K = 20$ in. Using the Pythagorean theorem, students find AA' as follows:

$$\begin{aligned}
(AK)^2 + (A'K)^2 &= (AA')^2,\\
12^2 + 20^2 &= (AA')^2,\\
144 + 400 &= (AA')^2,\\
544 &= (AA')^2,\\
AA' &= \sqrt{544} \approx 23.3.
\end{aligned}$$

Thus, the second method uses 23.3 in. of ribbon, 8.7 in. less than the first method.

Have students find the amount of ribbon needed, using both methods, for boxes with the same volume as the box on the student page but with different dimensions. For example, a cube 4 in. on a side also has a volume of 64 in.^{3} The first method again requires 32 in. of ribbon, but the second method needs only 22.6 in. If the box is 1 in. deep and its length and width are each 8 in., its volume is still 64 in.^{3} The first method uses 36 in. of ribbon and the second method uses 25.5 in. A store that wraps a lot of packages finds this a very large difference in the amount of ribbon it must use.

Extension

This Extension provides an excellent algebra review. To find the width of a box wrapped using the first method, students must set up and solve the equation:

$$\begin{aligned}
4(2) + 2(10) + 2w &= 44,\\
8 + 20 + 2w &= 44,\\
2w &= 16,\\
w &= 8.
\end{aligned}$$

Thus, the dimensions of the box are 2 in., 10 in., and 8 in.

Students must use the Pythagorean theorem in the second method:

$$\begin{aligned}
(2 + 10 + 2 + 10)^2 + (w + 2 + w + 2)^2 &= (44)^2,\\
(24)^2 + (2w + 4)^2 &= (44)^2,\\
576 + 4w^2 + 16w + 16 &= 1936,\\
4w^2 + 16w - 1344 &= 0,\\
w^2 + 4w - 336 &= 0.
\end{aligned}$$

Using the quadratic formula,

$$\begin{aligned}
w &= \frac{-4 \pm \sqrt{16 + 1344}}{2}\\
&= \frac{-4 \pm \sqrt{1360}}{2}\\
&= -2 \pm 2\sqrt{85}\\
&= 16.4 \quad \text{or} \quad -20.4.
\end{aligned}$$

Because the box cannot have a negative length, the dimensions of the box using this method are 2 in., 10 in., and 16.4 in.

Regular Polyhedra

You probably remember that a polyhedron is a solid bounded by plane polygons. The one shown below is a hexahedron.

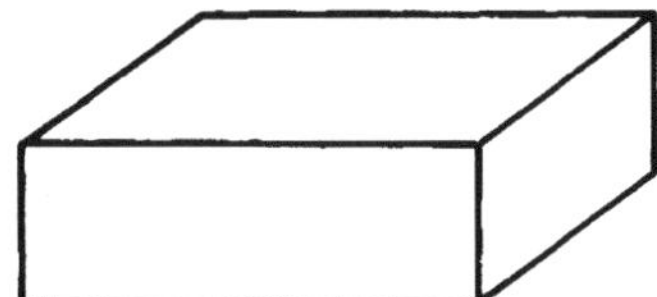

How many faces does it have? ________ How many edges? ________ How many vertices? ____

In a *regular polyhedron*, all the faces are congruent regular polygons and the same number of polygons meet at each vertex. How many regular polygons are there? ________ How many regular polyhedra do you think there are? ______

First consider regular polyhedra with equilateral triangles as faces. There must be at least three polygons at a vertex of a polyhedron or it won't be a solid. Suppose there are six equilateral triangles at a vertex. What will the sum of the measures of the angles at a vertex be? _____________ Is it possible to have a regular polyhedron with six equilateral triangles at a vertex? _____________ How many equilateral triangles *can* there be at a vertex of a regular polyhedron? ________ How many squares can there be at a vertex of a regular polyhedron? _____________ How many regular pentagons *can* there be? ________ Are there any other regular polygons that can form the faces of a regular polyhedron? ________ The five regular polyhedra are

Tetrahedron
Hexahedron (Cube)
Octahedron
Dodecahedron
Icosahedron

There is another way to show there can only be five regular polyhedra. What is the formula for the measure of an angle of a regular polygon of n sides? _____________ If there are p regular polygons at each vertex of the polyhedron, what is the sum of the angle measures at each vertex? _____________ Because this sum must be less than 360°,

$$p\left[\frac{(n-2)180}{n}\right] < 360 \quad \text{or} \quad (p-2)(n-2) < 4.$$

Both p and n must be greater than 2 because a polyhedron must have more than two faces at a vertex and a polygon must have more than two sides. Complete the following table for the only possible values of p and n:

p	n	$(p-2)(n-2)$	Name of Polyhedron	Number of Faces	Number of Vertices	Number of Edges
3	3					
3	4					
4	3					
3	5					
5	3					

Look at the last three columns of the table. Write a formula that relates the number of faces (F), vertices (V), and edges (E) of regular polyhedra: ______ ____________ Check to see if your formula works for nonregular polyhedra.

EXTENSION! A regular tetrahedron can be inscribed in a larger regular tetrahedron. Each vertex of the inscribed one is at the center of each face of the larger one. Using the preceding table, determine which regular polyhedra can be inscribed in other regular polyhedra.

Teacher's Notes for Regular Polyhedra

One of the unfortunate casualties of the decline of solid geometry's importance in the curriculum has been the consideration of regular polyhedra. These aesthetically pleasing figures are motivational by themselves. More importantly, they illustrate many basic geometric concepts. This activity introduces the regular polyhedra and their properties and presents a complex, yet satisfyingly clean proof that only five regular polyhedra can be formed. Moreover, Euler's theorem is intuitively developed from the discussion of the regular polyhedra.

NCTM Standards

1	2	3	4	5	6	7	8	9	10
•	•	•			•	•		•	•

Presenting the Activity

Models of the five regular polyhedra are very useful in presenting the activity. They are available commercially, but it's also easy to make them using heavy paper and enlargements of the patterns at the end of these notes. It's also helpful to have models of various nonregular polyhedra available.

There are 6 faces, 12 edges, and 8 vertices of the hexahedron shown. Point out that the quadrilateral faces need not be rectangles; many combinations of trapezoids and parallelograms will also work.

Because there are an infinite number of regular polygons, some students will think there are also an infinite number of regular polyhedra. Most of this activity is devoted to showing there can be only five regular polyhedra. Students should be familiar with the definitions of a face, edge, and vertex of a polyhedron. If not, review them briefly using the figure at the top of the student page. Emphasize that a regular polyhedron must have congruent faces and must have the same number of edges meeting at each vertex.

If six equilateral triangles meet at a vertex, the sum of the measures of the angles at the vertex will be 360°. Thus, the vertex will "flatten out" and all six triangles will be in the same plane. There can only be three, four, or five equilateral triangles at a vertex of a regular polyhedron. Similarly, there can be only three squares or three pentagons. If a regular polygon has more than five sides, its vertex angle will be 120° or greater. Thus, the sum of three vertex angles will be 360° or greater and the polygon cannot be a face of a regular polyhedron. Thus, there are only five ways to form regular polyhedra.

The next part of the activity gives a more formal proof that only five regular polyhedra exist. The measure of each angle of a regular polygon of n sides is

$$\frac{(n-2)180}{n}.$$

If there are p regular polygons at each vertex of the polyhedron, the sum of the angle measures at each vertex is

$$p\left[\frac{(n-2)180}{n}\right].$$

Thus,

$$p\left[\frac{(n-2)180}{n}\right] < 360,$$

$$180p(n-2) < 360n,$$

$$p(n-2) < 2n,$$

$$pn - 2p - 2n < 0,$$

$$pn - 2p - 2n + 4 < 4,$$

$$p(n-2) - 2(n-2) < 4,$$

$$(p-2)(n-2) < 4.$$

Students can use the figures on the student page or models of the regular polyhedra to complete the table. The completed table is

p	n	$(p-2)(n-2)$	Name of Polyhedron	Number of Faces	Number of Vertices	Number of Edges
3	3	1	Tetrahedron	4	4	6
3	4	2	Hexahedron	6	8	12
4	3	2	Octahedron	8	6	12
3	5	3	Dodecahedron	12	20	30
5	3	3	Icosahedron	20	12	30

Ask students why the values of p and n given in the table are the only possible values they can have. [As noted on the student page, $p > 2$ and $n > 2$. In addition, p and n must be whole numbers and the product $(p-2)(n-2)$ must be less than 4.]

The last three columns in the table give the formula $F + V = E + 2$. This is Euler's theorem and is true for all polyhedra. Have students test the theorem on various other polyhedra.

Extension

Because a hexahedron has six faces, its inscribed polyhedron must have six vertices, one at the center of each face. Thus, an octahedron can be inscribed in a hexahedron. Similarly, a hexahedron can be inscribed in an octahedron. In the same way, a dodecahedron can be inscribed in an icosahedron and an icosahedron in a dodecahedron.

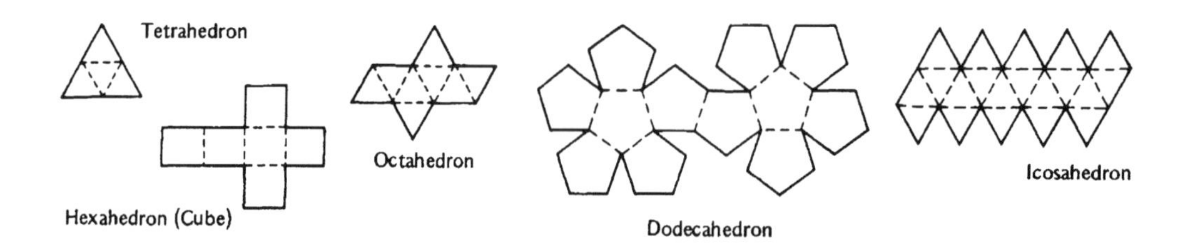

Cavalieri's Principle

Suppose you have two neat stacks of pennies, 20 in each. Then you push one stack to distort it as shown. How do the volumes of the two stacks compare?

Now consider the following two solids:

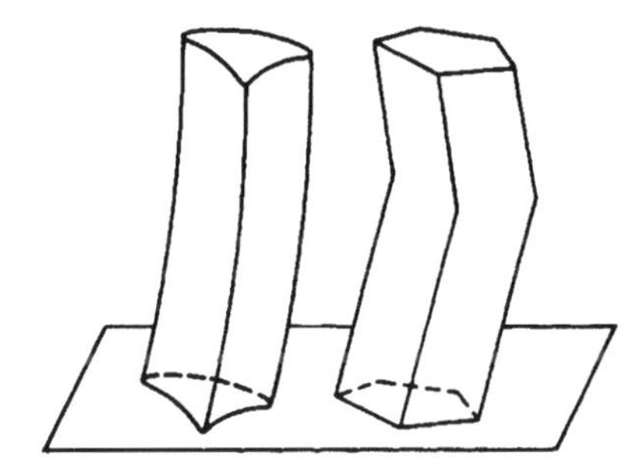

The solids have the same altitude and their bases lie in the same plane. When a plane parallel to their bases is passed through both solids, their cross sections always have the same area. How do the volumes of the two solids compare?

The relationship above is called *Cavalieri's principle* and is stated as:

> Consider two solids and a plane. Suppose every plane parallel to the given plane that intersects one of the solids also intersects the other solid and the resulting cross sections have the same area. Then the two solids have the same volume.

Cavalieri's principle seems pretty obvious—no big deal about it. Surprisingly, it can be used to find the volume of a sphere. The following figure shows a sphere with radius r and a cylinder with radius r and altitude $2r$.

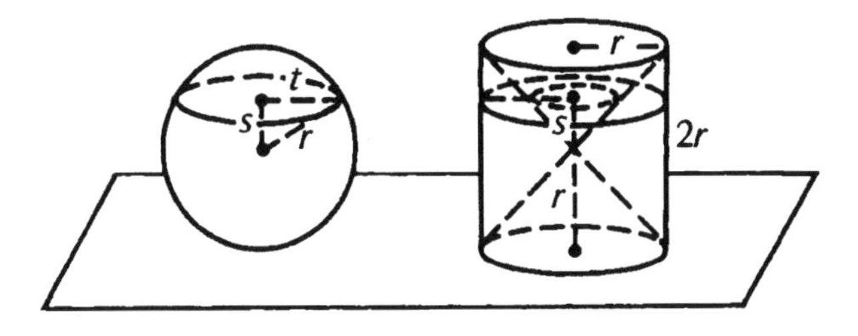

The cylinder has two cones removed; each cone has radius r and altitude r. A plane parallel to the given plane is passed through both solids. Let's consider the crosssectional areas. The cross section of the sphere is a circle with radius t. What is the area of this circle? $A =$ ________. What is the relationship between t, r, and s? ____

What is the area of the circle in terms of r and s? $A =$ ____________________.

The cross section of the cylinder with the cone removed is a ring. What is the radius of the larger circle? ________. What is the radius of the smaller circle? ________ What is the area of the ring? $A =$ ______________. From Cavalieri's principle, what is true about the volumes of the two solids? ________________

____

Using the formulas for the volume of a cylinder and a cone, what is the volume of the sphere? ____

EXTENSION! What is the relationship among the volumes of a cylinder, the inscribed sphere of the cylinder, and the inscribed cone of the cylinder?

Teacher's Notes for Cavalieri's Principle

Many mathematical formulas are presented to high school students with no explanation of why they are true. Thus, students often find it difficult to remember the formulas. The formula for the volume of a sphere is a good example of this. This formula is usually presented to students without proof. However, with Cavalieri's principle, the derivation of this formula is quite easy. Another advantage of Cavalieri's principle is that it is graphically obvious. Students are thus easily able to grasp why the formula for the volume of a sphere is as it is.

NCTM Standards

1	2	3	4	5	6	7	8	9	10
•	•	•	•		•	•	•	•	•

Presenting the Activity

Students should easily see that the volumes of the two stacks of pennies are equal because the volume of every penny is the same. You may want to have students consider volumes of other solids made from stacks of $3'' \times 5''$ note cards.

The volumes of the next two solids on the student page are also equal. This leads directly to Cavalieri's principle. Although Cavalieri's principle can be proved, it must be considered as a postulate in high school geometry because the proof requires calculus.

The cross-sectional area of the sphere should be easy for students to visualize. The area of this circle is πt^2. By the Pythagorean theorem, $t^2 + s^2 = r^2$, so $t^2 = r^2 - s^2$. By substituting for t^2, the area of the circle is $\pi r^2 - \pi s^2$.

The cross section of the cylinder with the cone removed may be more difficult for students to visualize. The following figure may help.

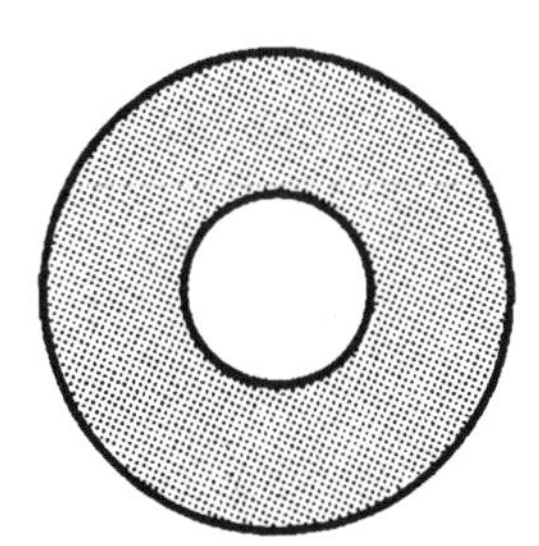

The radius of the larger circle is r, the radius of the cylinder. The radius of the smaller circle is the radius of the cone at a distance of s from the vertex of the cone. Point out that the altitude of the cone is the same length as its radius. Thus, the radius of the cone with altitude s will be s. So the radius of the smaller circle is s and the area of the ring is $\pi r^2 - \pi s^2$.

Therefore, the cross-sectional areas are equal and, by Cavalieri's principle, the two solids have the same volume. The volume of the cylinder is $2r \cdot \pi r^2$ and the volume of each cone is $\frac{1}{3}\pi r^2 \cdot r$. Thus,

$$\begin{aligned}
\text{volume of sphere} &= 2r \cdot \pi r^2 - 2 \cdot \frac{1}{3}\pi r^2 \cdot r \\
&= 2\pi r^3 - \frac{2}{3}\pi r^3 \\
&= \frac{4}{3}\pi r^3.
\end{aligned}$$

Extension

All your students should be able to do this Extension. They should use figures similar to those that follows and the dimensions given on the figures:

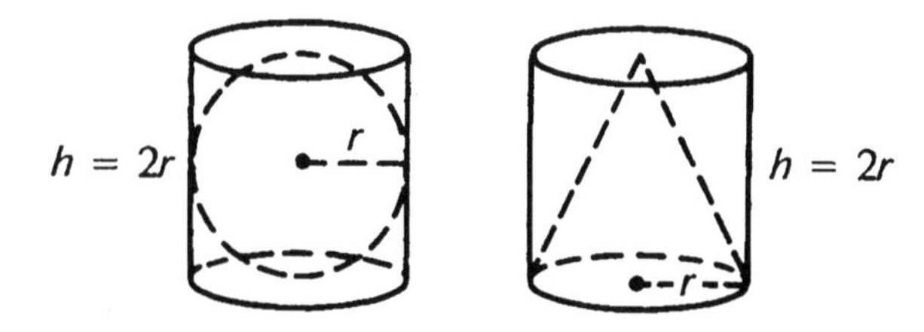

Point out that a sphere inscribed in a cylinder will touch the sides, top, and bottom of the cylinder. Similarly, a cone inscribed in a cylinder will have the same base as the bottom of the cylinder and the vertex will touch the top of the cylinder. The volumes of the three solids will be

$$\begin{aligned}
\text{volume of cylinder} &= 2r \cdot \pi r^2 \\
&= 2\pi r^3, \\
\text{volume of sphere} &= \frac{4}{3}\pi r^3, \\
\text{volume of cone} &= \frac{1}{3}\pi r^2 \cdot 2r \\
&= \frac{2}{3}\pi r^3.
\end{aligned}$$

Therefore, the volume of the cylinder is equal to the volume of the sphere plus the volume of the cone.

The Jolly Green Giant?

Can a giant 60 ft tall exist? Why can a fly walk on water, but be helpless when covered with water? The answers to these questions depend on the volume and surface area of the giant and the fly. First, consider the surface area and volume of a solid. If the distance between every pair of points in a solid is multiplied by n, what is the surface area of the solid multiplied by? ____________ What is the volume multiplied by? ____________

Now compare a 60-ft giant to a 6-ft man. The giant is not only 10 times as tall as the man, he is also 10 times as wide and 10 times as thick. The weight of the giant depends on his volume. If the man weighs 200 lb, what does the giant weigh? ____________ Each time the man takes a step, 200 lb are supported by his leg bone. If the cross-sectional area of the man's leg bone is 1 in.^{2}, what is the cross-sectional area of the giant's leg bone? ____________ When the giant takes a step, how many pounds are supported by each square inch of his leg bone? ____________ What do you think will happen to the giant when he takes a step? ____

____

How is an animal such as a hippopotamus able to support its weight? ______

____

____

Now consider the fly. The weight, volume, and surface area for a person and a fly are shown in the following table.

	Weight	Volume	Surface Area	Water
Person	150 lb	5000 in.^{3}	1500 in.^{2}	
Fly	0.00002 lb	0.0006 in.^{3}	0.1 in.^{2}	

A film of water covering an object is about 0.02 in. thick. Find the number of cubic inches of water it would take to cover the person and the fly.

One cubic inch of water weighs about 0.04 lb. How many pounds of water are covering the person? ____________ How many pounds are covering the fly? ____________ Compare the weight of the person to the weight of the water covering him and the weight of the fly to the weight of the water covering it. Why is a fly helpless when covered with water? ________________________

____

EXTENSION! A mouse falls down a 1000-yd mine shaft and is able to get up and walk away. A cat chasing the mouse also falls down the shaft. The cat is killed. Why?

Teacher's Notes for The Jolly Green Giant?

The ratio of surface area to volume is one of the most important—if not the most important—determinants in the evolution of all animal life forms. It doesn't just affect structural stability and locomotion; it affects the more basic processes of respiration and metabolism. Thus, this activity offers a good pragmatic application and should be an especially attractive assignment for students who feel geometry is of interest only to mathematicians and civil engineers.

NCTM Standards

1	2	3	4	5	6	7	8	9	10
•	•	•	•		•	•	•	•	

Presenting the Activity

There are many biological applications of the volume and surface area of solids. J. B. S. Haldane presented a variety of them in a delightful essay, "On Being the Right Size," in *The World of Mathematics* (J. R. Newman, ed., NY: Simon & Schuster, 1956, pp. 952–957). These additional applications can easily be discussed along with this activity.

If the distance between every pair of points in a solid is multiplied by n, the surface area is multiplied by n^2 and the volume by n^3. Using a cube with a side of a given length, students can easily verify this.

The weight of the giant depends on his volume, so his weight will be 1000 times the man's weight or 200,000 lb. The cross-sectional area of the giant's leg bone is only 100 times that of the man or 100 in.^{2}. Thus, for the giant, 2000 lb are supported on each square inch of leg bone, whereas only 200 lb are supported on the man's leg bone. If the giant takes a step, he breaks his leg. Our giant has little to be jolly about.

Discuss how increased size has affected some professional athletes. Professional basketball players today are much taller and heavier than their counterparts two generations ago. During a game, the stress on a player's feet and legs is sometimes enough to snap bones.

Some animals' legs have adapted in a way that has enabled them to become very large. For example, the rhinoceros and hippopotamus have short, thick legs. Thus, every pound of weight still has enough bone cross section to support the animal. A hippopotamus also lives primarily in water, and this helps to support its weight. Whales have adapted entirely to water and so can attain even larger size.

A fly weighs such a small amount that it is supported by water. However, when a fly is covered by water, what happens to it depends on its surface area rather than its volume. The person in the table will be covered by 30 in.^{3} of water, and the fly will be covered by 0.002 in.^{3} (1500 in.^{2} $\times$ 0.02 in. $=$ 30 in.^{3}; 0.1 in.^{2} $\times$ 0.02 in. $=$ 0.002 in.^{3}). The person will be carrying 1.2 lb of water ($30 \times 0.04 = 1.2$). This is only 0.008 of his weight ($1.2 \div 150 = 0.008$). The fly is, of course, carrying less water, only $0.002 \times 0.04 = 0.00008$ lb. However, this is 4 times its body weight. Compare this to the 150-lb person carrying 600 lb of water. Thus, a fly's life is in danger every time it tries to get a drink of water.

Note: In the foregoing discussion, the phenomenon of surface tension has not been introduced. This is actually the more serious threat to the insect in water, but it doesn't bear on the concept we're trying to teach and would produce more confusion than clarification.

Extension

Again, this is a matter of the ratio of area to weight or, more precisely, mass. If the cat and mouse were falling in a vacuum, then both would accelerate at the same rate and both would go splat at the bottom of the mine shaft. However, in a resistive medium such as air, the retarding force is directly proportional to the cross-sectional area of the falling object—the principle on which parachutes operate. The cat's cross section is about 10 or 20 times that of the mouse, but its weight is 100 or 200 times as great. Thus, when the cat hits the bottom of the mine shaft, it is traveling at a much higher velocity. You may also want to point out to your students that the mouse survives the ordeal only if it gets out of the way before the cat hits.

CHAPTER 6

Geometric Applications

- Mathematics on a Billiard Table ♦
- Bypassing an Inaccessible Region
- The Inaccessible Angle
- Minimizing Distances ⋆
- Problem Solving—A Reverse Strategy

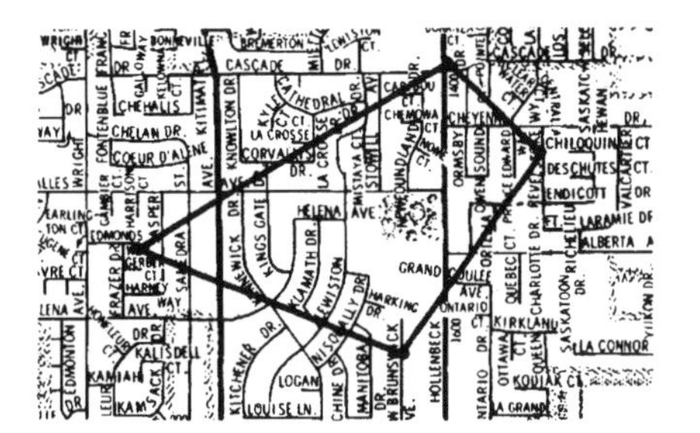

Most students need some evidence that what they're studying has some value outside the classroom. Several activities in this book are an attempt to fill this need: "Taxicab Geometry," "Projective Geometry," "This Wraps It Up," and "The Jolly Green Giant?" are good examples. Unfortunately, most high-school geometry students have not yet studied physics or chemistry, and so many of the practical applications of geometry are difficult to present. Nonetheless, an applications section is necessary, and the activities in this category have been found to be among the most popular.

"Mathematics on a Billiard Table" is one of the most highly motivating activities in this volume because it appeals to such a wide range of students. Moreover, this activity can be presented early in the course; students need to have learned only the properties of perpendicular bisectors and isosceles triangles.

"Bypassing an Inaccessible Region" and "The Inaccessible Angle" both deal with artificial, yet somewhat baffling constructions using a straightedge and a compass. In each case, students are asked to solve "real" problems that require some fairly creative thinking. Both of the problems have several correct answers, allowing different students to approach the problem in different ways. Both lead students through one solution to the problem and then challenge them to find another solution. Placement of these two activities depends on when basic constructions are studied.

"Minimizing Distances" includes an application of the equiangular point discovered in "Napoleon's Theorem." Thus, the two activities are best used together over a 2-day period. In this way, students can immediately apply a geometric concept to a real-world problem. If students have studied "Taxicab Geometry," the equiangular point can be compared to the minimum distance point in taxicab geometry. In "Minimizing Distances," distance as the crow flies is important and driving distance is not.

The proofs given in the early part of geometry courses, such as proving two triangles congruent, often give students some trouble. With diligence, however, they can often get through these proofs by sheer repetition. However, when more difficult proofs are introduced, it is obvious that many students do not understand how to approach writing proofs. "Problem Solving—A Reverse Strategy" provides students with some tools for understanding proofs. The activity shows students how to approach a proof and analyze it in reverse. It is strongly recommended that this activity be presented immediately after parallel lines and the triangle congruence theorems. This strategy should be emphasized in presenting all proofs and problems. Students' proof-writing and problem-solving skills will improve dramatically when they approach a problem by first analyzing what they want to find or prove.

Mathematics on a Billiard Table

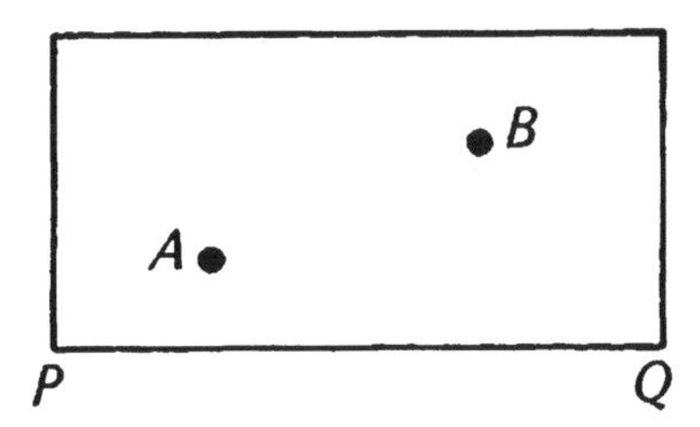

The preceding figure is a top view of a billiard table. Suppose you want to rebound ball A from cushion $\overline{PQ}$ so that A will then hit ball B. Mark a point R where you think ball A should hit on $\overline{PQ}$ for this to happen.

It turns out that point R should be located so that the path from A to R to B is the *shortest* path from A to $\overline{PQ}$ to B. You know that the shortest path between two points is a straight line, but that doesn't seem to apply here—or does it? Look at the following figure:

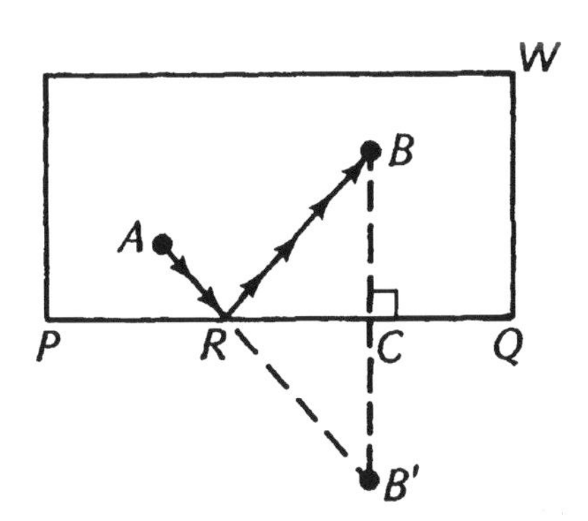

$\overline{BB'}$ is constructed perpendicular to $\overline{PQ}$ at C and $\overline{BC} \cong \overline{B'C}$. B' is called the *reflected image* of B in $\overline{PQ}$. Then $\overline{AB'}$ is drawn intersecting $\overline{PQ}$ at R.

Is $\overline{ARB'}$ the shortest path between A and B'? ____________________

Is A to R to B' the same length as A to R to B? ____________________

Prove that $AR + RB$ is the shortest path by selecting any other point S on $\overline{PQ}$ and showing that $AS + SB > AR + RB$. (Use $\triangle ASB'$.) ____________________

____

____

Now consider a billiard shot where ball A rebounds off cushions $\overline{PQ}$ and $\overline{WQ}$ before hitting ball B. This time reflect B into $\overline{WQ}$ to get image B'; then reflect B' into $\overleftrightarrow{PQ}$ to get B''. Where $\overline{AB''}$ intersects $\overleftrightarrow{PQ}$ will determine R and where $\overline{RB'}$ intersects $\overline{WQ}$ will determine T.

Thus, the desired path of ball A is $\overline{AR} \longrightarrow \overline{RT} \longrightarrow \overline{TB}$. Locate R and T on the following figure:

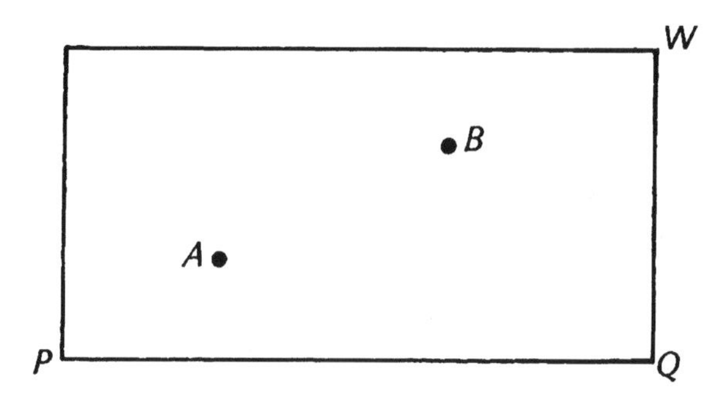

Find another way to locate R and T by first reflecting A in $\overline{PQ}$.

EXTENSION! Find points R, T, and V on $\overline{PQ}$, $\overline{WQ}$, and $\overline{SN}$, respectively, which will allow ball A to rebound off cushions $\overline{PQ}$, $\overline{WQ}$, and $\overline{SN}$ (in order) before hitting ball B.

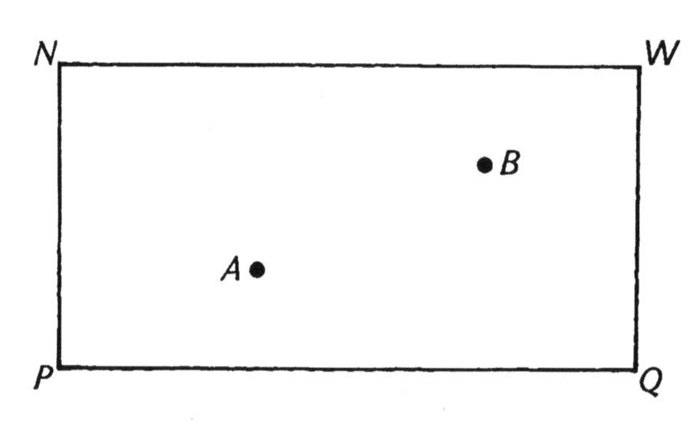

Teacher's Notes for Mathematics on a Billard Table

Many mathematicians like pool and billiards because the paths of the balls can be calculated mathematically. This activity shows students how these paths are calculated. The only prerequisites are the basic constructions and the properties of perpendicular bisectors.

NCTM Standards

1	2	3	4	5	6	7	8	9	10
		•	•		•	•		•	•

Presenting the Activity

There may be some disagreement among students as to the location of point R. If there are any experienced pool or billiard players in the class, have them explain how they would locate R. Most pool players know that R is the point where $m\angle ARP = m\angle BRQ$. However, they would usually not realize that this is the *shortest* path from A to $\overline{PQ}$ to B.

Next students are shown how to locate R using geometric constructions. Whereas $\overline{BB'} \perp \overline{PQ}$ and $\overline{BC} \cong \overline{B'C}$, $\overline{PQ}$ is the perpendicular bisector of $\overline{BB'}$. Thus, $RB = RB'$ and $AR + RB = AR + RB'$.

Now discuss the measures of angles ARP and BRQ. Whereas $\overline{RB} \cong \overline{RB'}$, $\triangle RBB'$ is isosceles and $m\angle BRQ = m\angle B'RQ$. Also $\angle ARP$ and $\angle B'RQ$ are vertical angles, so $m\angle B'RQ = m\angle ARP$. It follows that $m\angle ARP = m\angle BRQ$, which should be gratifying to the pool players in the class.

To prove that $AR + RB$ is the shortest path, students should select S on $\overline{PQ}$ and draw $\overline{AS}$, $\overline{SB}$, and $\overline{SB'}$ as shown in the following diagram. (*Note*: S can be on either side of R.)

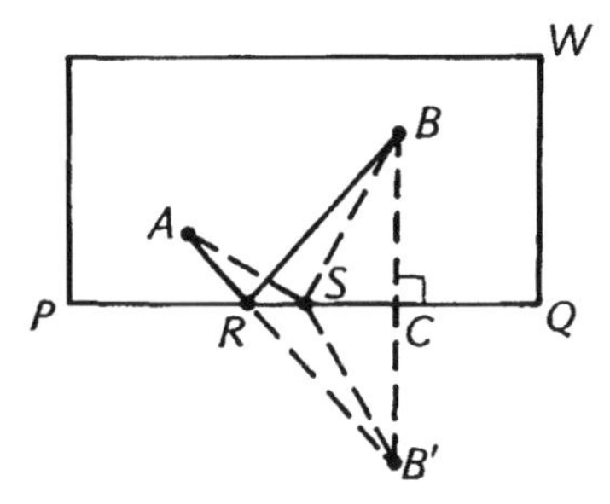

Because $\overline{PQ}$ is the perpendicular bisector of $\overline{BB'}$, $\overline{RB} \cong \overline{RB'}$ and $\overline{SB} \cong \overline{SB'}$. In $\triangle ASB'$, $AB' < AS + SB'$ (triangle inequality). Whereas $AB' = AR + RB'$,

$$AR + RB' < AS + SB',$$
$$AR + RB < AS + SB.$$

Students may be interested in how a mirror could be used to locate R: Place a mirror along the cushion $\overline{PQ}$ and, from position A, sight ball B in the mirror. This precisely locates point R.

In practice the physical billiard balls do not rebound from the rails exactly as light does. The balls come off the rail a little "flat." That is, the angle of reflection, $\angle BRQ$, is slightly smaller than the angle of incidence, $\angle ARP$, with the percentage difference between the two angles becoming greater as $\angle ARP$ becomes smaller. (As $\angle ARP$ approaches 90°, the difference approaches zero.) Nevertheless, the path of light is a good approximation, and the diamonds or markers imbedded in the rails are there to assist in just this sort of calculation.

Now students should consider the problem of rebounding from two cushions ($\overline{PQ}$, then $\overline{QW}$) before hitting B. The completed construction is shown:

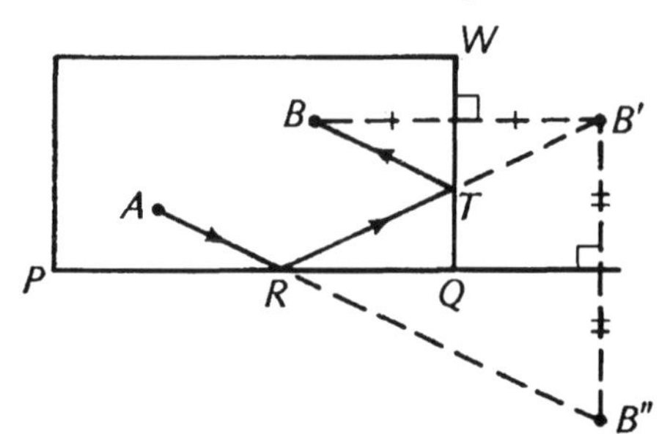

You may wish to consider again how mirrors could be used to locate R and T. Sighting from A into mirror $\overline{PQ}$, two images of B will appear. The first is the previously found B', a direct reflection from B. The second image (which, in this case, will appear a few inches to the left of B', as seen by the viewer) is a reflection of B's reflection in $\overline{QW}$.

The alternate construction would produce the following figure:

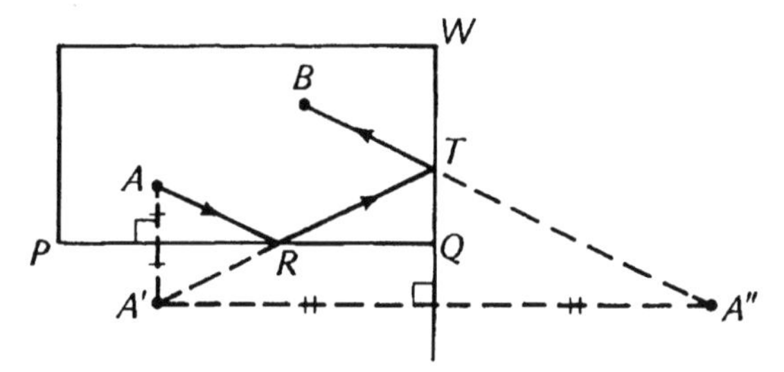

Extension

Points R, T, and V are located as shown in this the following construction:

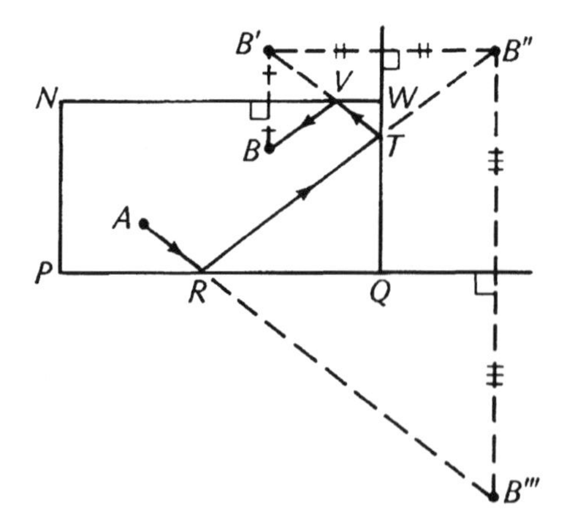

Bypassing an Inaccessible Region

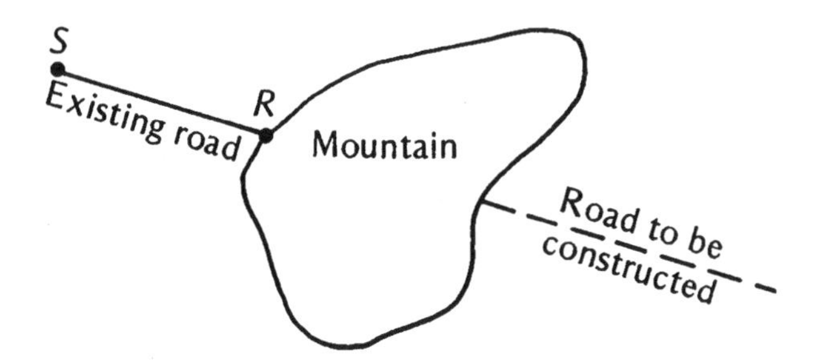

Two towns are separated by a mountain. For several years the townspeople have been planning to build a road connecting the towns with a tunnel through the mountain. They have been working in stages and plan to construct the tunnel last. So far, they have constructed a road from one town (at point S in the accompanying figure) to the mountain (point R). As the next step, they plan to construct a road from the other side of the mountain to the second town. They want this road to be collinear with the existing road. Suppose you want to find the path for this new road using only straightedge and compass constructions. In addition, suppose you can't touch or reach over the mountain. How would you construct the path of the new road? ______________________________

____

Let's consider one way to solve the problem. Using the figure

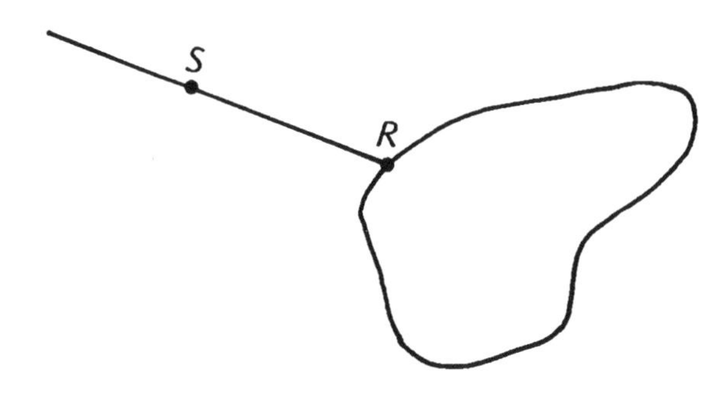

construct line l perpendicular to $\overline{SR}$ at any point N. Now construct line k perpendicular to l at point M. Be sure to choose M so that k will not touch the mountain. What is true about $\overline{SR}$ and k? ______________________________
Construct line t perpendicular to k at G. Choose G so that t and l are on different sides of the mountain. Find point H on t so that $GH = MN$. Finally, construct the line through H perpendicular to t. What kind of figure is $NMGH$?

____

Have you found the path for the new road? ______________________________

Another method for bypassing the inaccessible region uses an equilateral triangle. Using a separate sheet of paper, copy the mountain and $\overline{SR}$. Try to solve the problem using a construction that involves an equilateral triangle.

EXTENSION! Copy the inaccessible region in the accompanying diagram with points P and Q. Construct two collinear line segments, one with endpoint P and one with endpoint Q, extending in opposite directions from the inaccessible region. Remember you can use only a straightedge and a compass, and cannot touch or reach over the region.

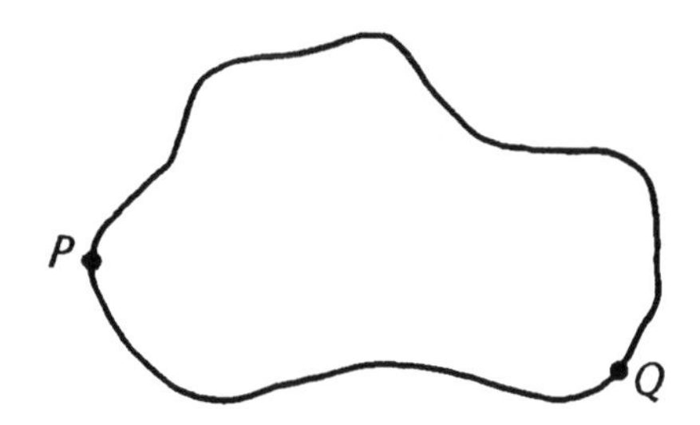

Teacher's Notes for Bypassing an Inaccessible Region

Most geometry texts present straightedge and compass constructions purely as an exercise and rarely use the constructions for problem solving. In this activity, students use only the basic constructions to solve a problem. However, to discover a correct method of solution requires some fairly creative thinking. Thus, the activity develops problem-solving skills as well as reinforcing construction techniques. An activity similar to this one is "The Inaccessible Angle."

NCTM Standards

1	2	3	4	5	6	7	8	9	10
		•	•		•	•			•

Presenting the Activity

After students have read through the problem and studied the figure, discuss how they might find a solution. Point out that because they can't go over the mountain, they must go around it. Review the basic constructions and encourage discussion of any solutions presented by the students. Then have the students work through the construction presented on the student page. The finished drawing should be similar to the following figure:

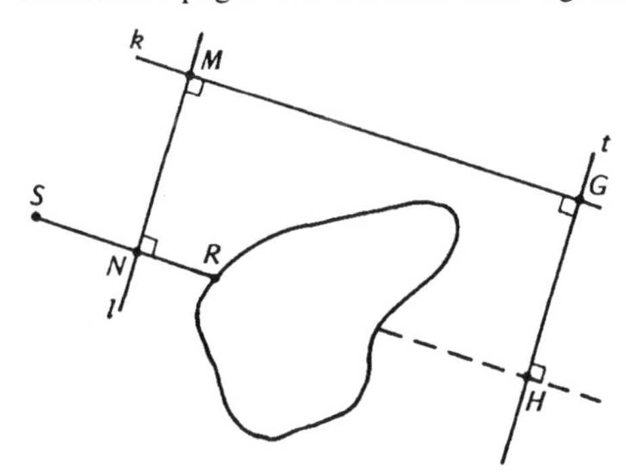

The students have constructed a rectangle, minus part of one side, so S, R, and H must be collinear.

The next method uses an equilateral triangle. Students simply construct an equilateral triangle as a separate figure. Then they can copy a 60° angle on $\overline{SR}$. The completed figure is

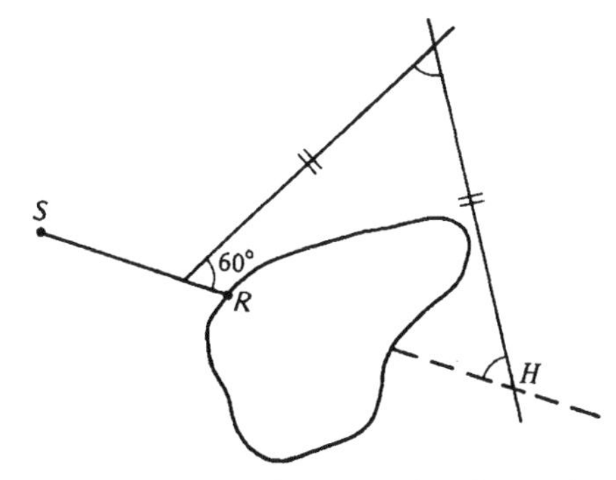

Students should see that any triangle can be used. They have only to construct a triangle around the mountain that is similar to the separately drawn triangle. Encourage students to think of other methods of solution that can be discussed and tested in class.

Extension

The problem of constructing a straight line through an inaccessible region when only the two endpoints are given is a much more challenging problem. Begin by drawing any convenient line segment from point P and construct a line perpendicular to it at a convenient point R, as shown:

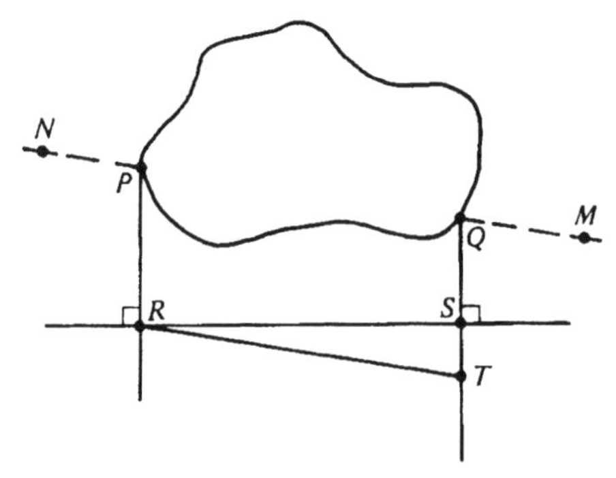

This perpendicular should not intersect the inaccessible region. Now construct a perpendicular from Q, to this last line drawn, intersecting it at S. Locate T on $\overleftrightarrow{QS}$ so that $PR = QT$. Draw $\overline{RT}$. Construct $\angle RPN \cong \angle PRT$ at P and construct $\angle TQM \cong \angle QTR$ at Q. This completes the required construction. $\overline{NP}$ and $\overline{QM}$ are extensions of side $\overline{PQ}$ of "parallelogram" $PRTQ$, and therefore are collinear.

There are many other methods for solving this problem. Many involve constructing similar triangles to then construct the two required lines. However students elect to approach this problem, they are apt to be led to a creative activity.

The Inaccessible Angle

Suppose you wanted to erect a wire antenna in an open field adjacent to a lake. The wire must be placed so that it will bisect an angle formed by two other wires. Unfortunately, the vertex of the angle formed by the two given wires is in the middle of the lake. Now let's make the problem even more difficult: You can use only a straightedge and pair of compasses, and cannot make any constructions in or over the lake. How can you construct the bisector of an angle you can't even touch? Study the following figure:

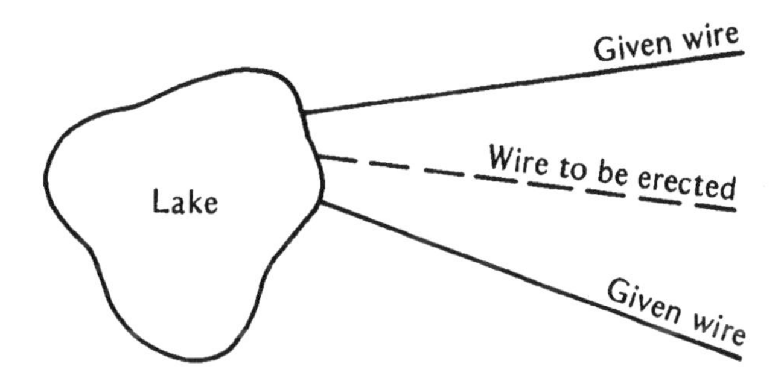

Can you think of a solution to the problem? ______________________________

____

____

There are actually many ways to solve this problem! Let's consider one of them. Begin by assuming the angle exists. Call the vertex of the inaccessible angle P, but draw no lines to or near it. Using the figure

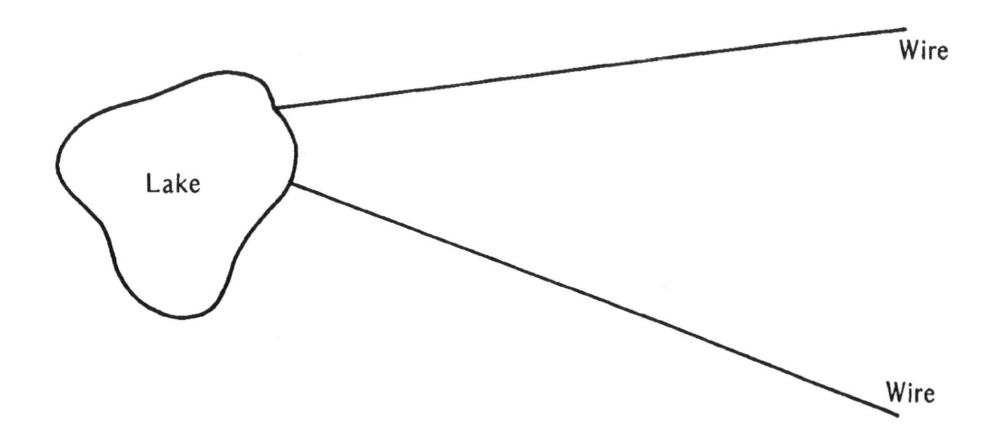

draw any line that intersects the two given rays at points A and B. Construct the bisectors of $\angle PAB$ and $\angle PBA$, and label their point of intersection M. Repeat this process for another line intersecting the given rays at C and D. Label the point of intersection of these two angle bisectors N. Draw line $\overleftrightarrow{MN}$.

Consider $\triangle ABP$. What must be true of the three angle bisectors of a triangle?

____

____

What point in $\triangle ABP$ is on the bisector of $\angle APB$? ____________

Now consider $\triangle CDP$. What point is on the bisector of $\angle CPD$? ________

Does $\overleftrightarrow{MN}$ bisect the inaccesible angle? ____________________

EXTENSION! The given method is just one way to construct the bisector of an inaccessible angle. Find another method.

Teacher's Notes for The Inaccessible Angle

Some mathematical problems have only one solution, so students are all required to think in the same way. The problem presented in this activity has many solutions and, so, it allows for different ways of thinking. Students have the opportunity to use the geometric relationships they have learned in new and creative ways. This type of creative thinking is essential in developing problem-solving skills.

Students should be familiar with the basic geometric constructions and the angle bisector concurrence theorem.

NCTM Standards

1	2	3	4	5	6	7	8	9	10
	•	•	•		•	•	•	•	•

Presenting the Activity

At first glance, the problem appears to be impossible. Assure students that this is not the case and encourage them to think of a solution. Then discuss the solution presented in the activity. The construction will produce a figure similar to

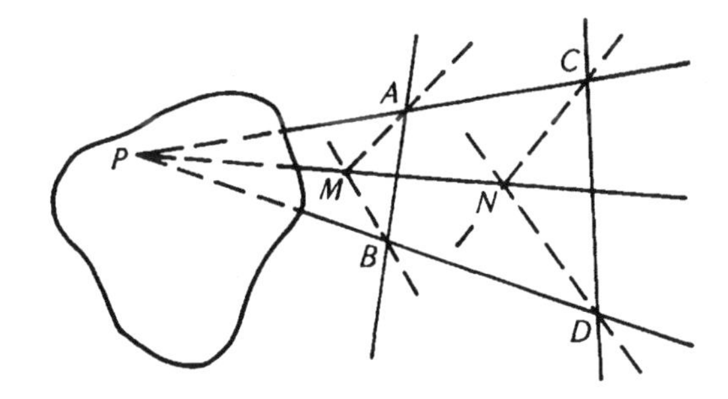

In $\triangle ABP$, the bisector of $\angle P$ must pass through M, because the angle bisectors of a triangle are concurrent. Similarly, in $\triangle CDP$, the bisector of $\angle P$ must pass through N. Thus, $\overleftrightarrow{MN}$ is the bisector of $\angle P$.

Extension

Two other solutions are presented next:

Solution 1. Begin by constructing a line parallel to one of the rays of the inaccessible angle. In the following figure, $\overleftrightarrow{RS}$ is parallel to $\overrightarrow{PT}$, and intersects $\overrightarrow{PQ}$ at point A.

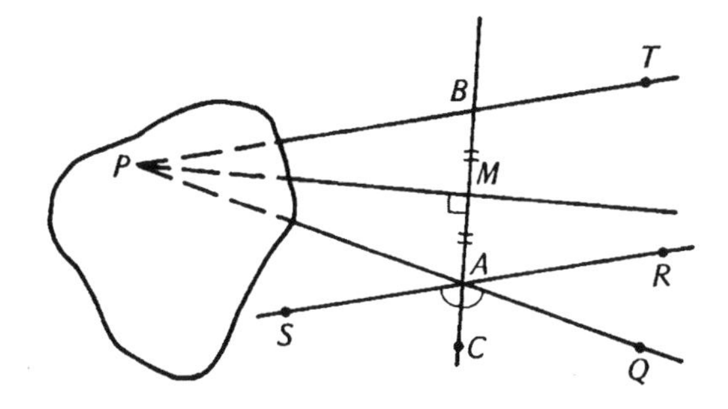

Construct the bisector of $\angle SAQ$, intersecting $\overrightarrow{PT}$ at B. Whereas $\overleftrightarrow{SR} \parallel \overrightarrow{PT}$, $\angle SAC \cong \angle PBA$. However, $\angle SAC \cong \angle CAQ \cong \angle PAB$. Therefore, $\angle PBA \cong \angle PAB$, and $\triangle PAB$ is isosceles. Because the perpendicular bisector of the base of an isosceles triangle also bisects the vertex angle, the perpendicular bisector of $\overline{AB}$ bisects the inaccessible angle.

Solution 2. Start by constructing a line $\overleftrightarrow{MN}$ parallel to one of the rays of the inaccessible angle, and intersecting the other ray at point A:

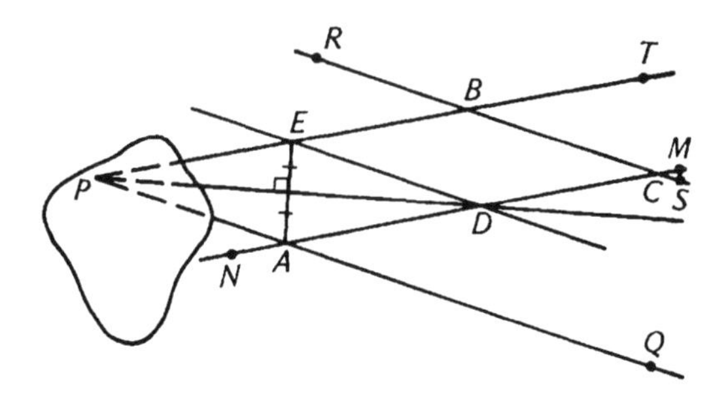

Then construct a line $\overleftrightarrow{RS}$ parallel to the other ray of the inaccessible angle, intersecting $\overrightarrow{PT}$ and $\overleftrightarrow{MN}$ at points B and C, respectively. Mark off a segment $\overline{AD}$ on $\overrightarrow{AC}$ that is the same length as $\overline{BC}$. Through D, construct $\overleftrightarrow{DE} \parallel \overrightarrow{PQ}$, where E is on $\overrightarrow{PT}$. Whereas $EBCD$ is a parallelogram and $ED = BC$, $ED = AD$. Because $PEDA$ is a parallelogram with two adjacent sides congruent ($\overline{ED} \cong \overline{AD}$), it is a rhombus. Thus, the diagonal $\overline{PD}$ is the bisector of the inaccessible angle. $\overline{PD}$ can be constructed simply by bisecting $\angle EDA$ or constructing the perpendicular bisector of $\overline{EA}$.

After presenting these solutions to your students, other solutions created by the students should follow directly. Free thinking should be encouraged to promote greater creativity.

Minimizing Distances

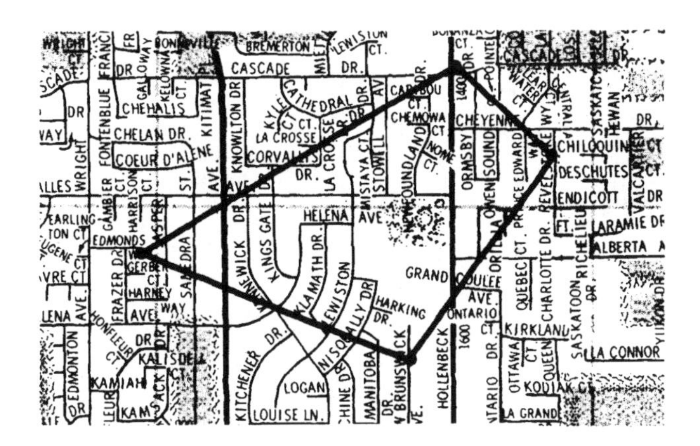

Four friends are planning to set up a special receiving unit to service their home alarm transmitters. Using a map of the town, they want to find a location for this receiver that makes the sum of the distances to each of their houses a minimum. The locations of the houses determine a quadrilateral as shown on the preceding map. Where do you think the transmitter should be located? ______________

____

A good point to try first as the *minimum distance point* is the point of intersection of the diagonals. Consider quadrilateral $ABCD$ with the diagonals intersecting at point Q.

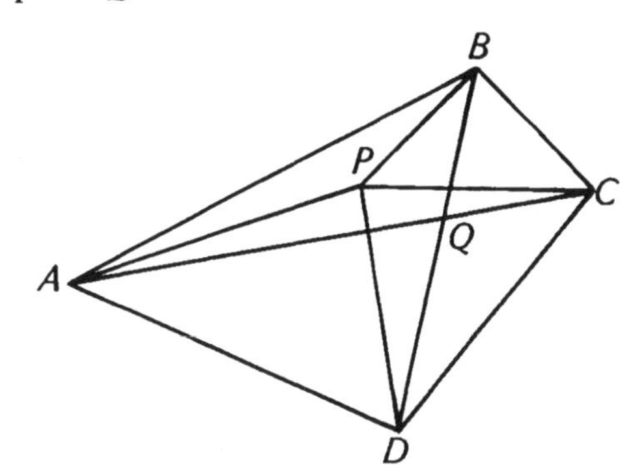

Now select *any* point P somewhere in the quadrilaterals, but not on the diagonals.

In $\triangle APC$, $PA + PC >$ _______ or $PA + PC >$ _______ + _______.

In $\triangle BPD$, $BP + PD >$ _______ or $BP + PD >$ _______ + _______.

Why is $QA + QB + QC + QD < PA + PB + PC + PD$? ______________

____

Is the point of intersection of the diagonals the minimum distance point in a quadrilateral? Why or why not? ______________________________

Suppose one of the four friends moves to another town. Now the locations of the houses determine a triangle.

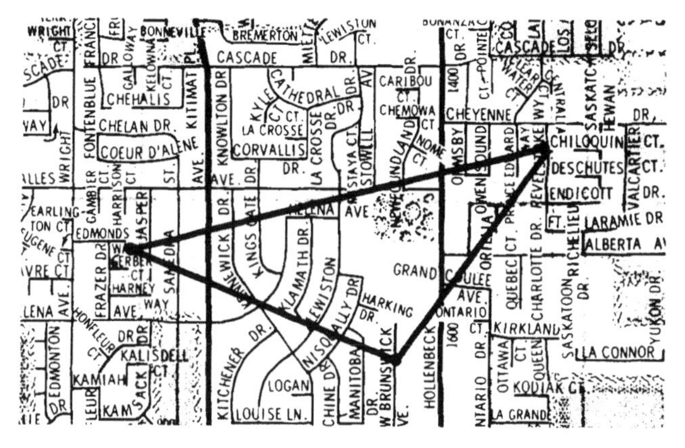

Where do you think the receiver should be located now? That is, where is the minimum distance point of a triangle? ______________________________

____

____

Surprisingly, the point we want is not one of the more familiar interior points of a triangle, such as the intersection of the medians or angle bisectors. It is the equiangular point or Napoleon point. How can you locate this point using a straightedge and compass? ______________________________

____

Is it possible to find the equiangular point of any triangle or should a restriction be placed on the angles of the triangle? ____________ Explain your answer.

____

EXTENSION! The sum of the distances from any point in the interior of an equilateral triangle to the sides of the triangle is constant. Use this to prove that the Napoleon point is the minimum distance point of a triangle.

Teacher's Notes for Minimizing Distances

This activity is an extension of "Napoleon's Theorem" and should follow it immediately. The equiangular point developed in "Napoleon's Theorem" is proved in this activity to also be the minimum distance point of a triangle—the point for which the sum of the distances from the point to the vertices is a minimum.

NCTM Standards

1	2	3	4	5	6	7	8	9	10
	•	•	•		•	•	•	•	•

Presenting the Activity

You can expect most students to guess that the point of intersection of the diagonals of the quadrilateral is the minimum distance point. The proof is not difficult: $AP + PC > AC$ or $AP + PC > AQ + QC$ because the sum of the lengths of two sides of a triangle is greater than the length of the third side.

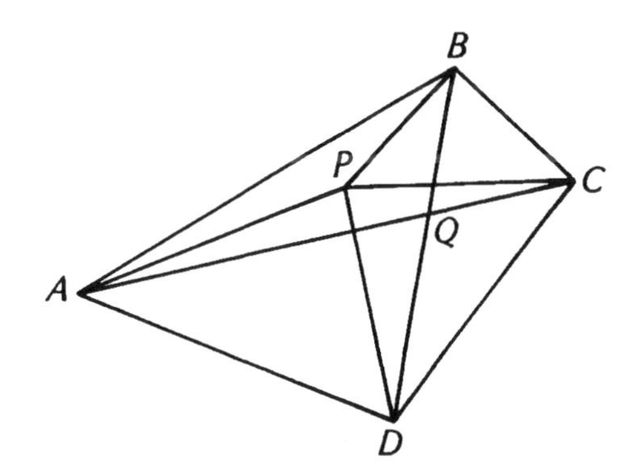

Similarly, $BP + PD > BQ + QD$. By addition, $PA + PB + PC + PD > QA + QB + QC + QD$.

Thus, the sum of the distances from the point of intersection of the diagonals of a quadrilateral to the vertices is less than the sum of the distances from any other interior point of the quadrilateral to the vertices.

The minimum distance point of a triangle is more difficult to locate. Have students locate the points of intersection of the medians, angle bisectors, and perpendicular bisectors of the sides. Then have them measure the distances from these points to the vertices of the triangle. They should also try any other points they might consider possibilities. Then have them consider the equiangular or Napoleon point. If "Napoleon's Theorem" has been presented recently, students should quickly recall how to locate this point. Students can see by measuring that the equiangular point is the minimum distance point of the triangle.

It is necessary to restrict the angles of the triangle to angles with measure less than 120°. If students don't understand why, have them try to locate the equiangular point in an obtuse triangle with one angle of measure 150°. The reason for the restriction should be obvious.

Extension

The proof that the Napoleon point is the minimum distance point uses the following figure.

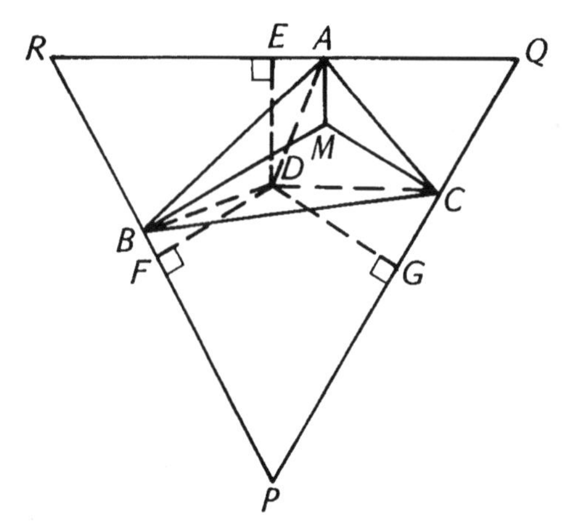

M is the point in the interior of $\triangle ABC$, where $m\angle AMB = m\angle BMC = m\angle AMC = 120°$. $\overline{RQ}$, $\overline{RP}$, and $\overline{QP}$ are drawn through A, B, and C perpendicular to $\overline{AM}$, $\overline{BM}$, and $\overline{CM}$, respectively. These lines form equilateral $\triangle PQR$. (To prove $\triangle PQR$ is equilateral, notice that each angle has measure $60°$. This can be shown by considering, for example, quadrilateral $AMBR$. Because $m\angle RAM = m\angle RBM = 90°$, and $m\angle AMB = 120°$, it follows that $m\angle ARB = 60°$.) Let D be any other point in the interior of $\triangle ABC$. We must show that the sum of the distances from M to the vertices is less than the sum of the distances from D to the vertices.

The sum of the distances from any point in the interior of an equilateral triangle to each of the sides is constant. Therefore, $MA + MB + MC = DE + DF + DG$ (where $\overline{DE}$, $\overline{DF}$, and $\overline{DG}$ are the perpendiculars to $\overline{RQ}$, $\overline{RP}$, and $\overline{QP}$, respectively). However, the shortest distance from an external point to a line is the length of the perpendicular segment from the point to the line. So, $DE + DF + DG < DA + DB + DC$. By substitution,

$$MA + MB + MC < DA + DB + DC.$$

If time permits, you may want to present a proof of the statement given in the Extension: The sum of the distances from any point in the interior of an equilateral triangle to the sides of the triangle is constant. One method of proving this uses the following figure.

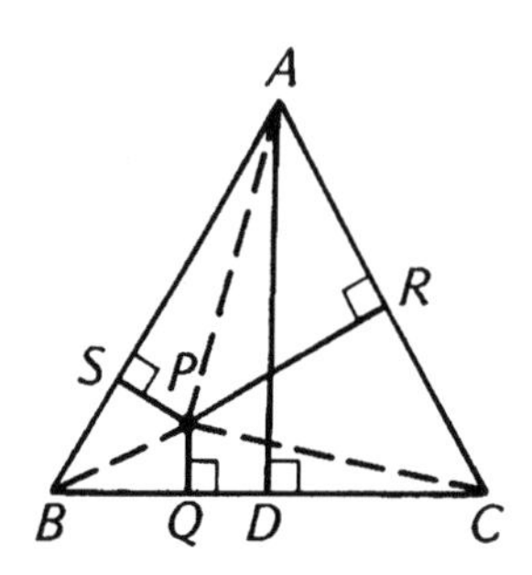

In equilateral $\triangle ABC$, $\overline{PR} \perp \overline{AC}$, $\overline{PQ} \perp \overline{BC}$, $\overline{PS} \perp \overline{AB}$, and $\overline{AD} \perp \overline{BC}$. Draw $\overline{PA}$, $\overline{PB}$, and $\overline{PC}$:

$$\begin{aligned}\text{area}\,\triangle ABC &= \text{area}\,\triangle APB + \text{area}\,\triangle BPC + \text{area}\,\triangle CPA \\ &= \frac{1}{2}(AB)(PS) + \frac{1}{2}(BC)(PQ) + \frac{1}{2}(AC)(PR).\end{aligned}$$

Whereas $AB = BC = AC$,

$$\text{area}\,\triangle ABC = \frac{1}{2}(BC)(PS + PQ + PR).$$

However, the area of $\triangle ABC = \frac{1}{2}(BC)(AD)$. Therefore, $PS + PQ + PR = AD$, a constant for the given triangle.

Problem Solving—A Reverse Strategy

How often have you looked at a completed geometry proof and thought, "That's really easy—once you know where to start." In many cases, the best place to start is at the end! Let's consider a simple algebra problem:

> If the sum of two numbers is 6 and the product of the same two numbers is 3, find the sum of the reciprocals of these two numbers.

To solve this problem, most people would use the two equations

$$x + y = 6,$$
$$xy = 3.$$

They would then solve for one of the variables and substitute in the other equation to find x and y. Then they would find the sum of the reciprocals. This method results in a lot of work. Try it and see.

Now see what happens if you work backward; that is, start out with the desired conclusion. You want to find the sum of the reciprocals of two numbers, $\frac{1}{x} + \frac{1}{y}$. What is the sum of these two fractions? ____________ What is the numerator and what is its value? ____________________________________

What is the denominator and what is its value? ____________________

What is the answer to the problem? ____________________________

This *reverse strategy* is the key to geometry proofs. Try it on the following proof.

Given: $\overline{AC} \parallel \overline{EF}$, $\overline{AC} \cong \overline{EF}$, and segment $\overline{BDCF}$ with $\overline{BD} \cong \overline{CF}$.

Prove: $\overline{AB} \parallel \overline{DE}$.

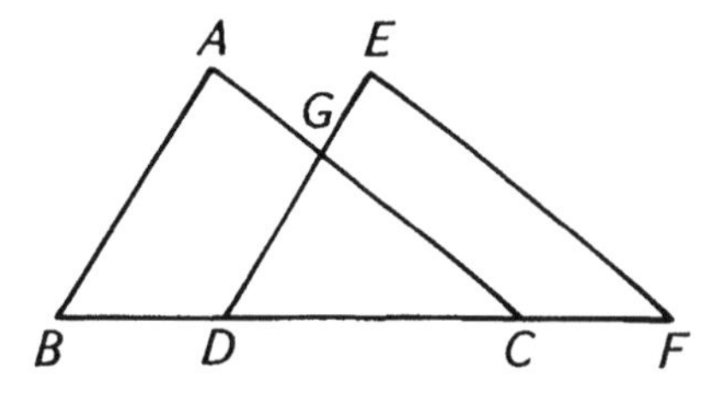

How can you prove lines parallel? ______________________________

____

Which line is a transversal of $\overline{AB}$ and $\overline{DE}$? ________ Is there a pair of corresponding angles or alternate interior angles for $\overline{AB}$ and $\overline{DE}$ using this transversal? If so, what are they? ____________ Which pair of triangles might you be able to prove congruent that include these two angles? ________________

Do you see where the reverse strategy is leading? Let's continue.

List some of the ways you can establish congruence between two triangles. ___

____

What congruent parts of triangles ABC and EDF are you given? __________

What information can you use to prove $\angle ACB \cong \angle EFD$? ______________

Why? ____

Using the remaining given information, which pair of sides can you prove congruent? _____________ Why? ____________________________________

What congruence theorem can you use to prove $\triangle ABC \cong \triangle EDF$? ________

____

Now the analysis is complete. Write the proof in the proper sequence.

EXTENSION! Given the following figure, with $\overline{AB} \cong \overline{AC}$ and $\overline{BE} \cong \overline{CE}$, prove that $\overline{DE} \cong \overline{FE}$.

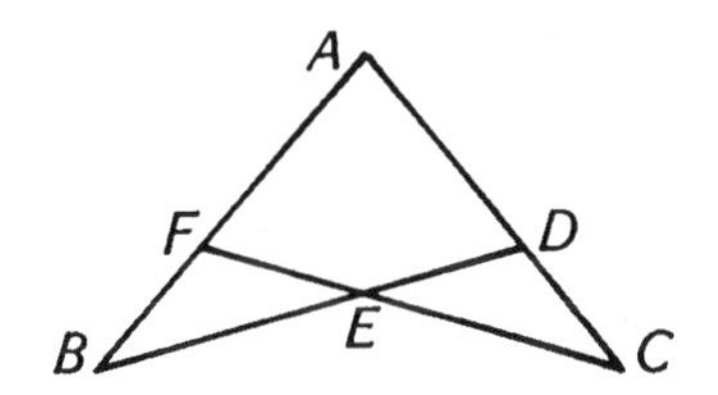

Teacher's Notes for Problem Solving—A Reverse Strategy

Most geometry students can write proofs once they have some idea where to begin. However, geometry texts usually teach proofs by example, rather than providing students with a definite strategy. This activity shows how to think through a proof by beginning at the end and working backward.

Students should be familiar with the congruence theorems and parallel lines. The activity should be presented as soon as possible after these concepts are learned, so students can begin using a reverse strategy immediately.

NCTM Standards

1	2	3	4	5	6	7	8	9	10
	•	•	•		•	•	•	•	•

Presenting the Activity

A reverse strategy is certainly not new. It was considered by Pappus of Alexandria about 320 A.D. In Book VII of Pappus' *Collection* there is a thorough description of the methods of analysis and synthesis. *Analysis* is the reverse strategy presented here—beginning with the desired conclusion and working backward until we reach something already known. Using *synthesis* reverses the process of analysis; it retraces the "reverse" steps and puts things in the order the proof requires.

The reverse approach to solving a problem becomes dramatically stronger, when the resulting solution becomes significantly more elegant. Students should work through the algebra problem on the student page using the first method discussed. By solving the first equation for y to get $y = 6 - x$, and then substituting into the second equation, they will get $x(6 - x) = 3$ or $x^2 - 6x + 3 = 0$. Then $x = 3 \pm \sqrt{6}$ and the two numbers are $3 + \sqrt{6}$ and $3 - \sqrt{6}$. Now the sum of their reciprocals is

$$\frac{1}{3+\sqrt{6}} + \frac{1}{3-\sqrt{6}} = \frac{(3-\sqrt{6}) + (3+\sqrt{6})}{(3+\sqrt{6}) \cdot (3-\sqrt{6})} = \frac{6}{3} = 2.$$

Now students should consider a reverse strategy. The sum of the fractions is $\frac{x+y}{xy}$. The two original equations reveal the values of the numerator and the denominator of the fraction. This produces the answer, $\frac{6}{3}$, immediately. It is obvious that for this particular problem, a reverse strategy is superior to the more common approach.

Analysis is essential to geometric proofs. Otherwise a student will consider the given information and proceed blindly, proving segments, angles, and triangles congruent until (if ever) the desired conclusion is reached. The next series of questions on the student page leads students through an analysis of a geometric proof. To prove lines parallel, students should realize they usually need congruent angles. By using $\overline{BF}$ as the transversal, $\angle ABC$ and $\angle EDF$ are a pair of corresponding angles. If $\angle ABC$ were congruent to $\angle EDF$, then $\overline{AB}$ would be parallel to $\overline{DE}$. Students can prove $\angle ABC \cong \angle EDF$ if they can prove $\triangle ABC \cong \triangle EDF$.

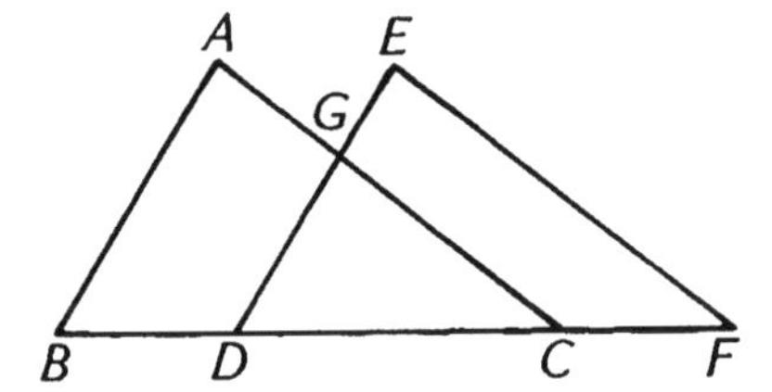

Students should begin to see where the reverse strategy is leading. Now they can prove segments and angles congruent, but with an aim in sight. They are given $\overline{AC} \cong \overline{EF}$, so they have one pair of congruent sides. In addition, $\angle ACB \cong \angle EFD$, because they are corresponding angles of parallel lines $\overline{AC}$ and $\overline{EF}$. Referring again to the given information, students see that $\overline{BD} \cong \overline{DF}$. Thus, students can use the side-angle-side congruence theorem to prove $\triangle ABC \cong \triangle EDF$. By retracing their steps in the opposite order, students should be able to write the proof in the proper sequence.

Extension

To prove segments congruent, students should realize they usually need congruent triangles. Thus, they must prove $\triangle BFE \cong \triangle CDE$. When they examine the given information, they see they already have two sides of these triangles congruent, $\overline{BE}$ and $\overline{CE}$. Whereas $\angle FEB$ and $\angle DEC$ are vertical angles, $\angle FEB \cong \angle DEC$. Now students should consider which congruence theorem to use. They can't use SAS, because then they would need $\overline{DE} \cong \overline{FE}$, and this is what they are trying to prove. The only possibility seems to be angle-angle-side, but which angles should they use?

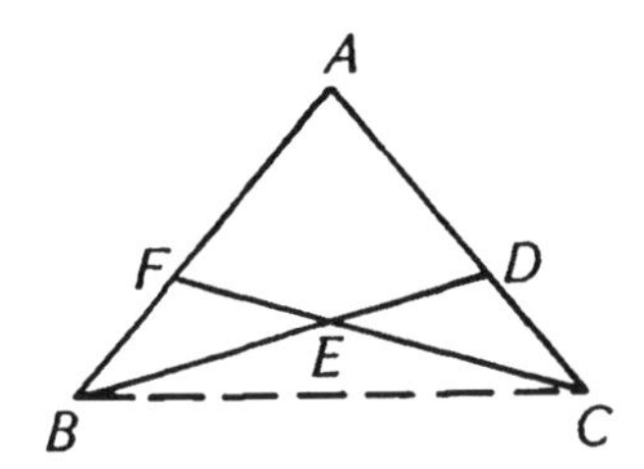

By referring to the given information, students will see that $\overline{AB} \cong \overline{AC}$. Thus, if segment $\overline{BC}$ is drawn, $\triangle ABC$ is isosceles and $\angle ABC \cong \angle ACB$. Similarly, $\triangle EBC$ is isosceles and $\angle EBC \cong \angle ECB$. Therefore, by subtraction, $\angle FBE \cong \angle DCE$, and $\triangle BFE$ can be proved congruent to $\triangle CDE$.

In his book *How to Solve It*, George Polya discussed a backward method of problem solving that is similar to the reverse strategy discussed in this activity. Polya emphasized the importance of the role of a teacher in presenting such methods to students when he states that "there is some sort of psychological repugnance to this reverse order which may prevent a quite able student from understanding the method it if is not presented carefully."

Further illustrations of the power of the reverse strategy can be found in *Problem-Solving Strategies for Efficient and Elegant Solutions* by A. S. Posamentier and S. Krulik (Corwin Press, 1998).

CHAPTER 7

Geometric Puzzlers

- Geometric Fallacies
- The Nine-Point Circle
- Equicircles ⋆
- More Equicircles ⋆
- Locus Methods ⋆

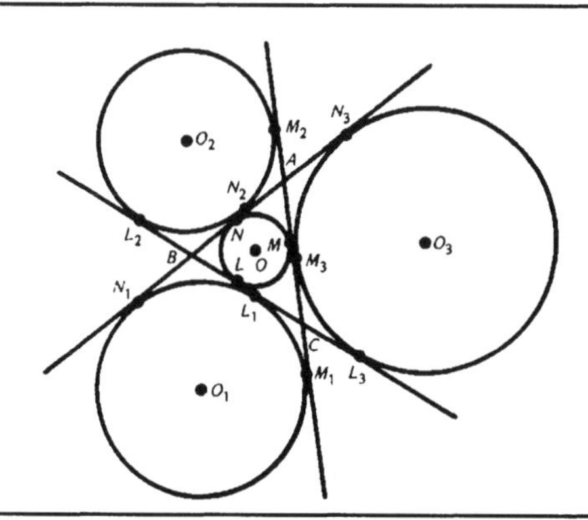

All five activities in this category will test the ingenuity of your students, but they are "puzzlers" for different reasons. The first presents two geometric paradoxes to unravel. The next three present geometric figures that lead to one relationship after another until many areas of geometric knowledge are tested. The last activity demonstrates how an apparently simple problem can become much more complicated as every possible case is considered.

The last four activities in this category must be presented late in the year because they draw on many areas of geometric knowledge. "Geometric Fallacies" can be presented after similar triangles have been studied. In geometry, fallacies are usually the result of incorrectly drawn figures; something that is not true appears to be so. Both of the fallacies in this activity are of this type and teach students to be careful in relying on figures. If you wish to present "Geometric Fallacies" before similar triangles are studied, have students draw the figure in the Extension on graph paper and in that way discover that it is not a triangle.

"The Nine-Point Circle" presents one of the most interesting figures in geometry, a circle with nine triangle-related points on it. These points are located by constructing the midpoints of the sides of the triangle and the three altitudes. The Extension relates the circumcenter, orthocenter, centroid, and center of the nine-point circle via the Euler line. Two class periods should be allowed for this activity so that verification for the construction can be presented.

The next two activities discuss the many relationships among the three escribed circles and the inscribed circle of a triangle. In "Equicircles," students find the lengths of various tangent segments in terms of the lengths of the sides of the triangle. In "More Equicircles," they write the radii of the equicircles in terms of the area and the lengths of the sides of the triangle. The two activities provide a comprehensive study that reinforces algebra and many concepts of geometry. The Extension for "Equicircles" asks students to generalize their results, an important scientific skill.

As in "Constructing Triangles," students' ingenuity and problem-solving skills are called upon in "Locus Methods." This activity leads students through a complete analysis of a construction problem by considering the locus of possible solutions. The ideas discussed in "Locus Methods" are fairly complex and in most cases the activity should be presented over at least two class sessions.

One Final Note

As the title of this series and several preceding statements indicate, these activities are meant to make mathematics come alive and be motivational—to kindle the interest—largely through the variety and novelty of problems that your students have not seen before. An interesting facet of motivation is how very contagious it is. Therefore, as a first cut, it is strongly recommended that you simply browse through this book and make a mental note of the activities that are especially appealing to you yourself. Then make the decision as to where in your course these activities best fit. Your own enthusiasm will assure a successful and enjoyable learning experience for your class.

Geometric Fallacies

Can we prove something true that isn't true? Will Rogers said that politicians do it all the time. It's a little tougher to do this in geometry than in politics, but read the following proof and see if you can find the fallacy.

To prove that a scalene triangle ($\triangle ABC$) is isosceles:

1. Draw the bisector of $\angle C$ and the perpendicular bisector of $\overline{AB}$. Their intersection is G. From G draw perpendiculars to $\overline{AC}$ and $\overline{CB}$. These meet the lines at points D and F, respectively (see Figure 1).

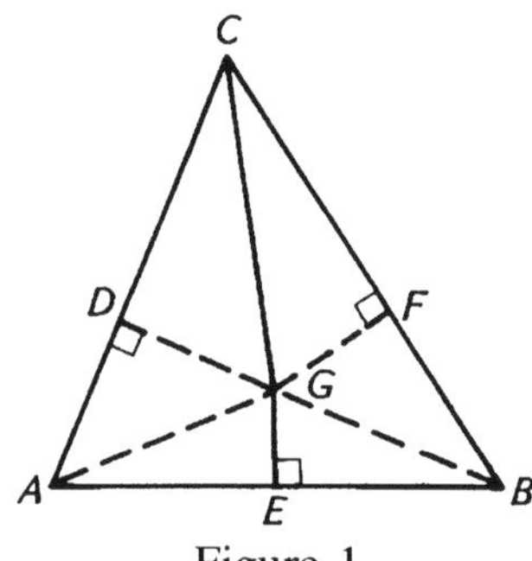

Figure 1

2. $\triangle CDG \cong \triangle CFG$ (angle-angle-side).
3. Therefore, $DG = FG$ and $CD = CF$ (corresponding sides of congruent triangles).
4. $AG = BG$ (G is a point on the perpendicular bisector of $\overline{AB}$).
5. $\triangle DAG \cong \triangle FBG$ (hypotenuse–leg).
6. Therefore, $DA = FB$ (corresponding sides of congruent triangles).
7. Thus, $AC = BC$ (addition).

What have you "proved"? ____________________________________

See if this proof will work with each of the following figures, where point G takes on various positions:

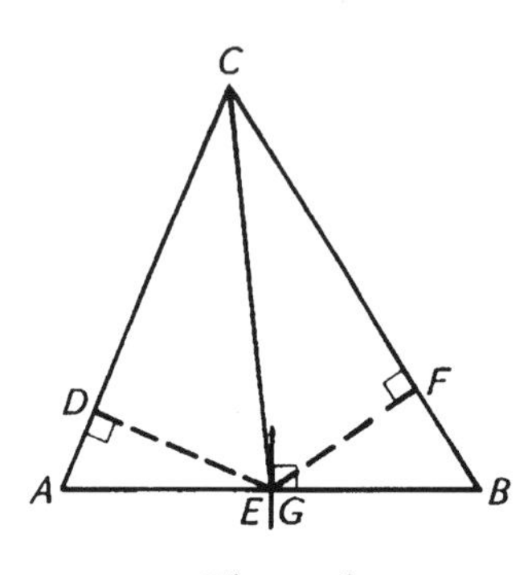

Figure 2

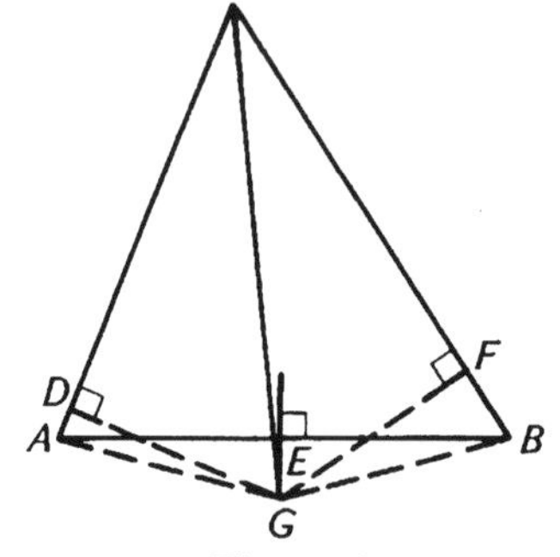

Figure 3

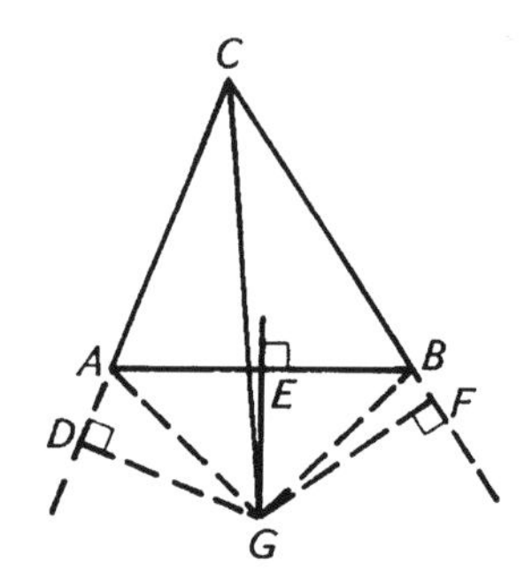

Figure 4

Where's the error in this "proof"? Using a straightedge and a pair of compasses, very carefully make the constructions indicated in step 1 of the "proof."

How does your diagram compare to Figures 1–4? ______________________

____

Where are points D and F? ______________________________________

What is wrong in the "proof"? ______________________________________

____

EXTENSION! The accompanying figure is made up of four right triangles, four rectangles, and a "hole." Find the sum of the areas of the right triangles and the rectangles. Then find the area of $\triangle PQM$. What is the area of the "hole?" Find the fallacy and, using geometric means, show why it is false.

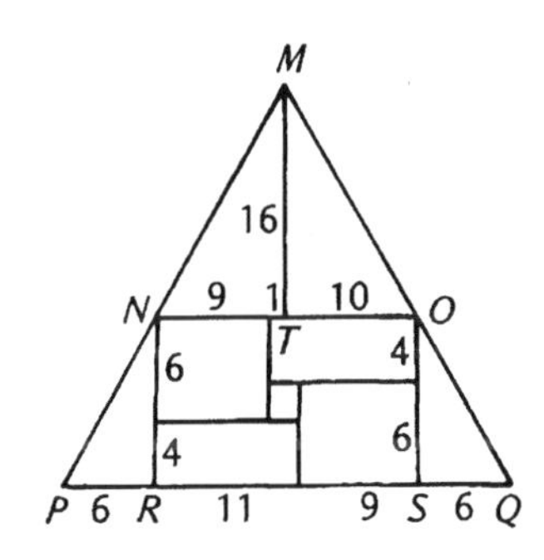

Teacher's Notes for Geometric Fallacies

When geometry students begin to write their own proofs, they may be careless in drawing the figure or, using some wishful thinking, they may rely on something that only appears to be true from the figure. This activity shows students how an error in reasoning can occur if a figure is not constructed accurately. You may want to compare this activity with "Algebraic Fallacies."

Students should know the various methods for proving triangles congruent. To work the Extension, they will need to be familiar with proofs of similar triangles.

NCTM Standards

1	2	3	4	5	6	7	8	9	10
	•	•	•		•	•	•	•	•

Presenting the Activity

The proof is a relatively easy one and many students will probably be convinced they have "proved" that a scalene triangle is isosceles. They will find the proof also works for Figures 2, 3, and 4. (The reason for step 7 using figure 4 is subtraction, rather than addition. Otherwise the proofs are identical.)

At this point students will be quite disturbed. They will wonder where the error was committed that permitted this fallacy to occur. Ask the students to consider whether *all* the possibilities for the figure are shown. Some students will realize that one perpendicular can be inside the triangle and the other outside. This possibility is not shown. Then ask your students to *carefully* make the constructions indicated in step 1. They will find a subtle error in the figures:

a. The point G *must* be outside the triangle.
b. When perpendiculars $\overline{DG}$ and $\overline{FG}$ meet the sides of the triangle, one will meet a side *between* the vertices, whereas the other will not.

The figure should look like this:

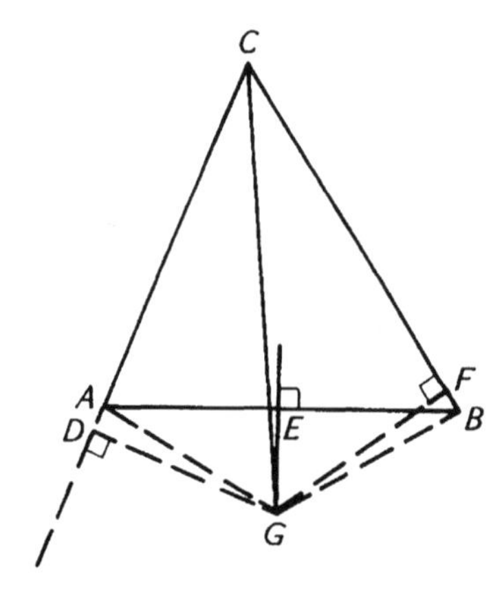

Thus, in step 7, the segments cannot be added or subtracted to get $AC = BC$. Some discussion of the concept of betweenness should follow.

To justify this conclusion (i.e., prove it) consider the circumcircle of $\triangle ABC$.

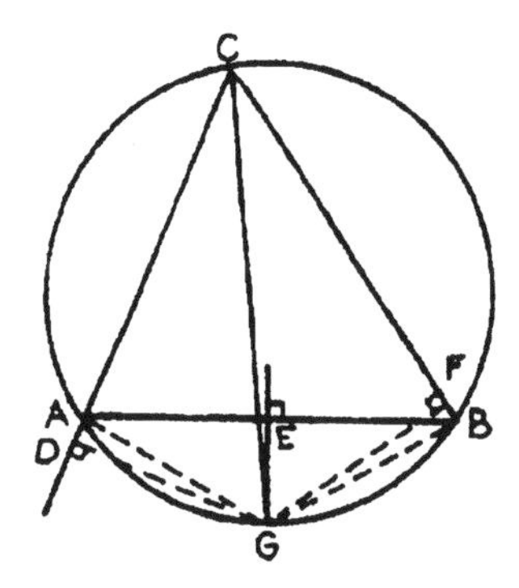

The bisector of $\angle ACB$ must contain the midpoint G of $\widehat{AB}$ (since $\angle ACG$ and $\angle BCG$ are congruent inscribed angles). The perpendicular bisector of $\overline{AB}$ must bisect $\widehat{AB}$, and therefore pass through G. Thus, the bisector of $\triangle ACB$ and the perpendicular bisector of $\overline{AB}$ intersect *outside* the triangle at G. This eliminates the possibilities illustrated in Figures 1 and 2.

Now consider inscribed quadrilateral $ACBG$. Since the opposite angles of an inscribed (or cyclic) quadrilateral are supplementary, $m\angle CAG + m\angle CBG = 180°$. If $\angle CAG$ and $\angle CBG$ are right angles, then $\overline{CG}$ would be a diameter and $\triangle ABC$ would be isosceles. Therefore since $\triangle ABC$ is scalene, $\angle CAG$ and $\angle CBG$ are not right angles. In this case one must be acute and the other obtuse. Suppose $\angle CBG$ is acute and $\angle CAG$ is obtuse. Then in $\triangle CGB$ the altitude on $\overline{CB}$ must be *inside* the triangle, while in obtuse $\triangle CAG$, the altitude on $\overline{AC}$ must be *outside* the triangle. (This is usually readily accepted without proof, but can be easily proved.) The fact that one and only one of the perpendiculars intersects a side of the triangle between the vertices destroys the fallacious "proof."

Extension

Students will find that the area of the eight regions (not including the "hole") is 416. Using the formula for the area of a triangle, the area of the entire figure is also 416. Students must determine how the area with and without the "hole" can be the same.

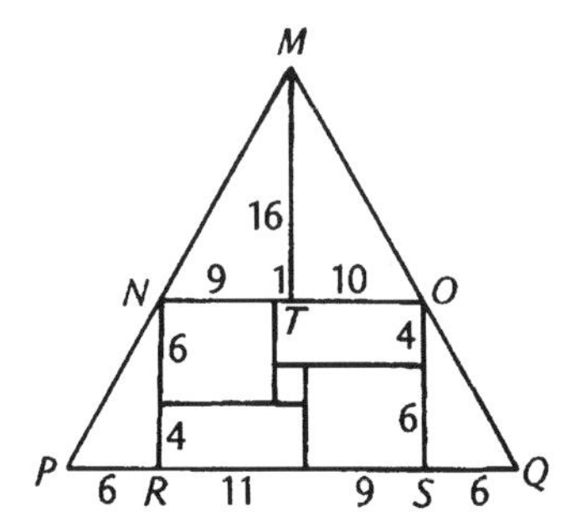

The fallacy occurs because the figure is *not* a triangle, because points M, N, and P are not collinear. If points M, N, and P were collinear, because $\angle RNO$ is a right angle, $\angle PNR$ is the complement of $\angle MNT$. Whereas $\angle NRP$ is a right angle, $\angle PNR$ is the complement of $\angle RPN$. Therefore, $\angle MNT \cong \angle RPN$ and $\triangle MNT \sim \triangle NPR$. However, this is not the case because the sides are not proportional. The same argument holds for points M, O, and Q. Therefore, the figure is a pentagon, and the formula we used to find the area is incorrect.

You may want to use the following books to present other geometric fallacies to your class.

Fallacies in Mathematics by E. A. Maxwell (Cambridge Univ. Press, Cambridge, UK, 1963).

Riddles in Mathematics by E. P. Northrop (Van Nostrand, New York, 1944).

Geometry, Its Elements and Structure (2nd ed.) by A. S. Posamentier, J. H. Banks, and R. L. Bannister (McGraw-Hill, New York, 1977, pp. 242–244, 270–271).

The Nine-Point Circle

Every triangle is *cyclic*. That is, there exists a circle that contains all of its vertices. So, by constructing a triangle, you have found three points through which exactly one circle can be drawn. These three points are *concyclic* and the circle could be called a three-point circle. How could you construct a four-point circle?

____

What must be true about a quadrilateral for it to be cyclic? _______________

____

____

Finding five concyclic points would be more difficult. Finding *nine* concyclic points seems impossible. However, there's a way to construct a nine-point circle that starts with a triangle. Using $\triangle ABC$ in the accompanying diagram,

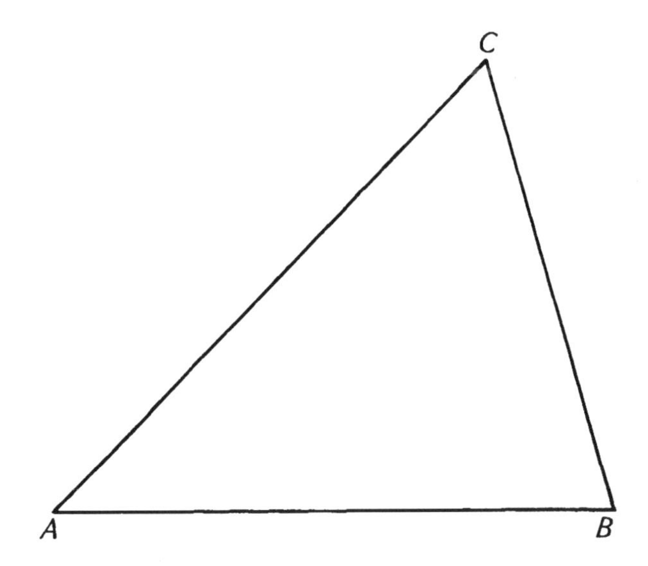

construct altitudes $\overline{AD}$, $\overline{BE}$, and $\overline{CF}$. Label their intersection, called the *orthocenter*, H. Now construct the midpoints of segments $\overline{AB}$, $\overline{BC}$, $\overline{AC}$, $\overline{AH}$, $\overline{BH}$, and $\overline{CH}$. Label these midpoints, respectively, U, V, W, X, Y, and Z. Points D, E, F, U, V, W, X, Y, and Z are concyclic and determine a nine-point circle. To draw the circle, you must find the center. Draw the segment that connects the midpoint of $\overline{CH}$ and the midpoint of $\overline{AB}$. Why is this segment the diameter of the nine-point circle? ____

____

____

Bisect the diameter and draw the circle.

Suppose $\triangle ABC$ is isosceles. How many points will you locate using the preceding method? _______ Why? ___________________________________

____

What happens if $\triangle ABC$ is equilateral? ___________________________

____

EXTENSION! On a separate sheet of paper, draw an acute, scalene triangle and construct its nine-point circle. Then construct the circumcircle of the triangle. Draw the line segment that connects the circumcenter and the orthocenter. This line is called the *Euler line*. Where is the center of the nine-point circle? How does the radius of the nine-point circle compare to the radius of the circumcircle? Draw the medians of the triangle. Where is their point of intersection?

Teacher's Notes for The Nine-Point Circle

The nine-point circle is in itself one of the most interesting figures in geometry. By using it to lead to a discussion of the Euler line in the Extension, students find a truly fascinating relationship between the circumcenter, the orthocenter, the centroid, and the center of the nine-point circle.

Although the material in the activity can be presented in one class period, you may wish to take two periods to thoroughly explore the topic and present verification for the construction of the nine-point circle. The activity should be presented late in a geometry course, because students must be familiar with the properties of circles, cyclic quadrilaterals, and concurrent segments in triangles.

NCTM Standards

1	2	3	4	5	6	7	8	9	10
	•	•	•		•	•	•	•	•

Presenting the Activity

Students know that any three points determine a circle. Thus, a triangle determines a three-point circle. To construct a four-point circle, students would construct a square (or a rectangle) and draw the diagonals. The intersection of the diagonals is the center of the circumscribed circle. Students should recall that a quadrilateral is cyclic if its opposite angles are supplementary. This fact is used in the subsequent justification of the construction of the nine-point circle.

Construction of the nine-point circle (shown next) should not be difficult for students.

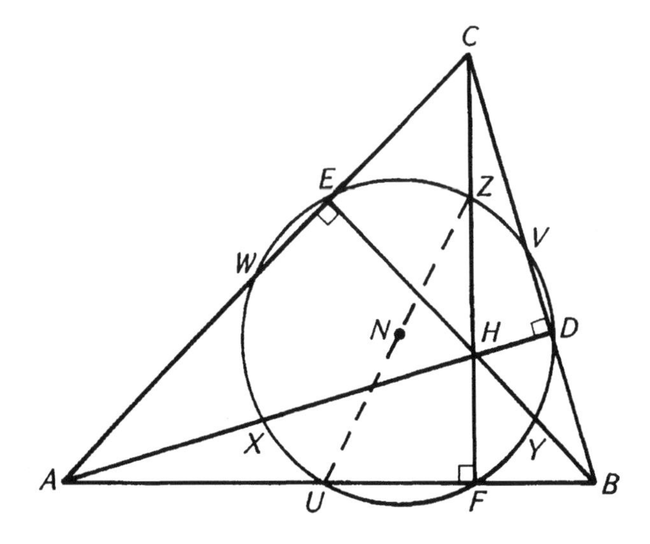

$\overline{UZ}$ is a diameter of the circle because $\angle ZFU$ is a right angle. Thus, it must be inscribed in a semicircle.

To prove the construction valid consider the figure

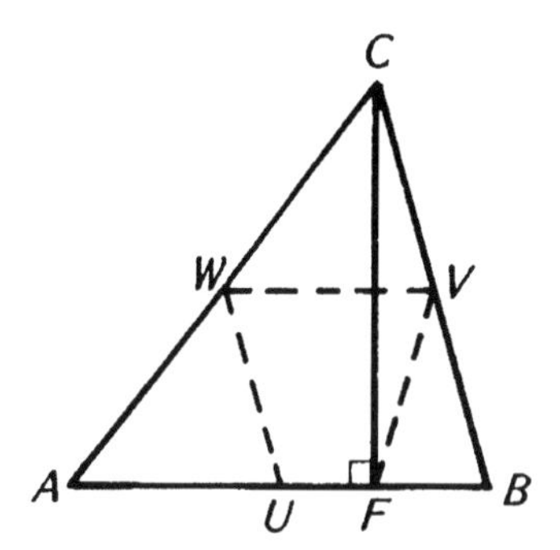

First prove $UFVW$ is an isosceles trapezoid: $\overline{WV}$ joins the midpoints of two sides of a triangle, so $\overline{WV} \parallel \overline{AB}$ and $UFVW$ is a trapezoid. $\overline{UW}$ joins the midpoints of $\overline{AB}$ and $\overline{AC}$, so $UW = \frac{1}{2}BC$. In right $\triangle BCF$, $\overline{FV}$ is the median to the hypotenuse, so $FV = \frac{1}{2}BC$. Therefore, $UW = FV$ and $UFVW$ is isosceles. Because an isosceles trapezoid is cyclic, F is on the circle determined by U, V, and W. Similarly, D and E are also on this circle.

Now consider this figure:

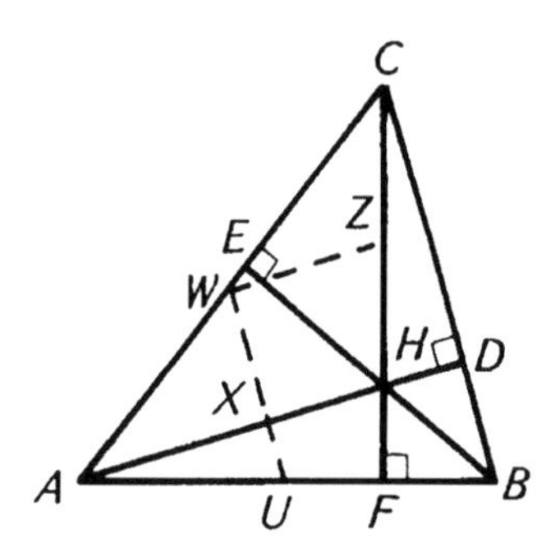

$\overline{WZ}$ joins the midpoints of two sides of $\triangle ACH$, so $\overline{WZ} \parallel \overline{AH}$. Similarly, in $\triangle ABC$, $\overline{WU} \parallel \overline{BC}$. $\overline{AD} \perp \overline{BC}$, so $\overline{WZ} \perp \overline{WU}$ and $m\angle ZWU = 90°$. Recall that $m\angle AFC = 90°$. Therefore, quadrilateral $ZWUF$ is cyclic, because its opposite angles are supplementary. This is the same circle established before, because three vertices (W, U, and F) are common with the six concyclic points. Similarly, X and Y are also on this circle.

If $\triangle ABC$ is isosceles, only eight points are found because the midpoint of the side opposite the vertex angle is the same point as the base of the altitude to that side. In an equilateral triangle, all three altitudes bisect the sides, so only six points are found. In this case, the center of the circle is at the orthocenter and the circle is inscribed in the triangle. Have students try the construction with an obtuse triangle where the orthocenter is outside the triangle.

Extension

The Euler line presents one of the most interesting relationships in geometry. (*Note*: Students will see the results easiest if the angles of the triangle are approximately 45°, 60°, and 75°.)

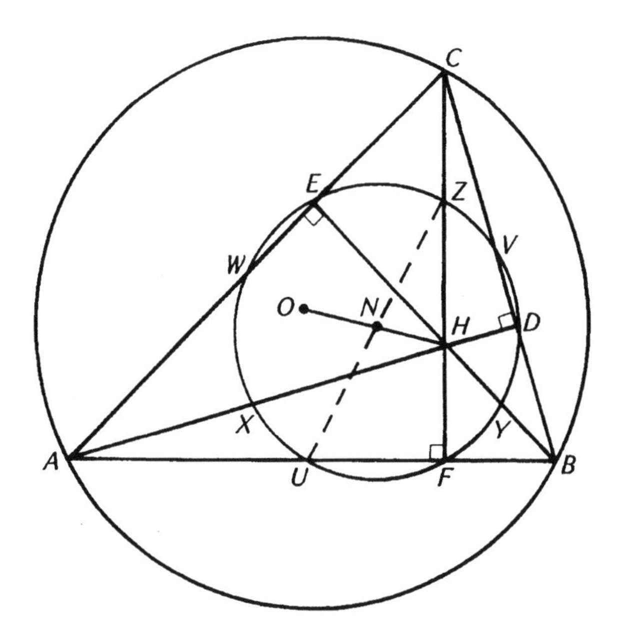

The Euler line in the preceding figure is $\overline{OH}$. N, the center of the nine-point circle, not only lies on the Euler line, but is also its midpoint. The radius of the nine-point circle is half the radius of the circumcircle. The intersection of the three medians (the centroid of the triangle) is also on the Euler line. The centroid trisects the Euler line.

Equicircles

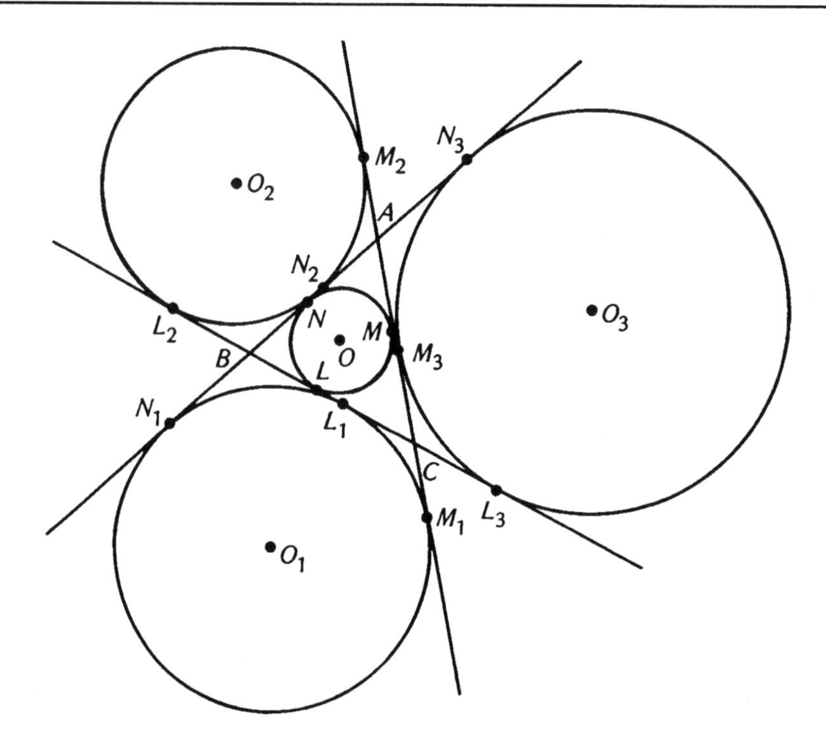

The preceding circles are the four *equicircles* of $\triangle ABC$. Circle O is the *inscribed* circle; the other three are the *escribed* circles of $\triangle ABC$. An escribed circle is outside the triangle and is tangent to all three of the lines that contain the sides of the triangle. Circle O is tangent at L, M, and N; circle O_1 is tangent at L_1, M_1, and N_1; circle O_2 is tangent at L_2, M_2, and N_2; and circle O_3 is tangent at L_3, M_3, and N_3.

Let's derive some of the relationships between a triangle and its equicircles. What is true about $\overline{AN_1}$ and $\overline{AM_1}$? ______________________

Why? ______________________

What segment has the same length as $\overline{BL_1}$ ________ As $\overline{CL_1}$? ________ The perimeter of $\triangle ABC = AB + BC + AC = AB + BL_1 + CL_1 + AC$.

What is the perimeter of $\triangle ABC$ in terms of AN_1? ____________ Thus, if s is the semiperimeter of $\triangle ABC$, $AN_1 = s$. Are there other segments with lengths equal to s? If so, what are they? ______________________

Many of the segments in the figure above can be written in terms of s, the semiperimeter, and the lengths of the sides of $\triangle ABC$, where $a = BC$, $b = AC$, and $c = AB$. First, let's find AN:

$$s = \frac{1}{2}(AN + BN + BL + CL + CM + AM).$$

We know $AM = AN$, $BL = BN$, and $CM = CL$, so,

$$
\begin{aligned}
s &= \frac{1}{2}(AN + BL + BL + CL + CL + AN) \\
&= \frac{1}{2}(2AN + 2BL + 2CL) \\
&= AN + BL + CL \\
&= AN + a, \\
AN &= s - a.
\end{aligned}
$$

Using the preceding method, what does BN equal? ____________________
Now find AN_2:

$$AN_2 = AM_2 = CM_2 - AC = \text{\_\_\_\_\_\_\_\_\_\_\_\_\_\_\_\_\_\_\_\_\_\_\_\_\_\_\_\_}.$$

Using the lengths of the sides of $\triangle ABC$ and AN_1, AN, BN, and AN_2 found previously, you can find the lengths of the tangent segments of various inscribed and escribed circles. Find each of the following lengths:

$N_1N_3 =$ ______________; $N_1N_2 =$ ______________;
$NN_1 =$ ______________; $NN_2 =$ ______________.

EXTENSION! Use your results from the preceding equations to write four general statements that describe the lengths of tangent segments in equicircles.

Teacher's Notes for Equicircles

Many, many relationships can be developed from the study of equicircles. This activity concentrates on tangent segments that are formed. ("More Equicircles" investigates properties of the radii of equicircles.) "Equicircles" reinforces identification of tangents and shows how algebraic methods can be used to solve geometric problems.

The Extension gives students practice in one of the most important processes in mathematics—writing general statements from equations. Students should be familiar with circles and tangents before this activity is presented.

NCTM Standards

1	2	3	4	5	6	7	8	9	10
•	•	•	•		•	•		•	•

Presenting the Activity

Students are already familiar with the inscribed circle of a triangle and the position of the escribed circles should be clear from the figure on the student page. The center of an escribed circle, an *excenter*, is the intersection of two exterior angle bisectors and one interior angle bisector. Excenter O_1, for example, is the intersection of the bisectors of interior angle BAC and exterior angles N_1BC and BCM_1.

The relationships derived in this activity are all based on (1) the theorem that two tangent segments from an external point to a circle are congruent and (2) algebraic manipulation. $AN_1 = AM_1$ by the mentioned theorem. Similarly, $BL_1 = BN_1$ and $CL_1 = CM_1$. By substitution in the formula for the perimeter,

$$\begin{aligned}\text{perimeter } \triangle ABC &= AB + BL_1 + CL_1 + AC \\ &= AB + BN_1 + CM_1 + AC \\ &= AN_1 + AM_1 \\ &= 2AN_1\end{aligned}$$

and AN_1 is one-half the perimeter of $\triangle ABC$. The other segments with lengths equal to s are $\overline{AM_1}$, $\overline{BL_3}$, $\overline{BN_3}$, $\overline{CL_2}$, and $\overline{CM_2}$.

The remainder of the activity develops the relationships between the lengths of tangent segments. To find the lengths of various tangent segments, students must first find the lengths of $\overline{AN}$, $\overline{BN}$, and $\overline{AN_2}$ in terms of a, b, c, and s. Using the method on the student page, BN is found as follows:

$$\begin{aligned}s &= \frac{1}{2}(AN + BN + BL + CL + CM + AM) \\ &= \frac{1}{2}(AM + BN + BN + CM + CM + AM) \\ &= \frac{1}{2}(2BN + 2CM + 2AM) \\ &= BN + CM + AM \\ &= BN + b,\end{aligned}$$

$$BN = s - b.$$

To find AN_2, students substitute $s = CM_2$ and $b = AC$. Thus, $AN_2 = s - b$.

Now students are ready to find the lengths of tangent segments in terms of the lengths of the sides of $\triangle ABC$. Ask students to describe each segment before they find its length.

$\overline{N_1N_3}$ is the common external tangent segment of two escribed circles:

$$\begin{aligned} N_1N_3 &= AN_1 + BN_3 - AB = s + s - c \\ &= 2s - c = a + b + c - c \\ &= a + b. \end{aligned}$$

$\overline{N_1N_2}$ is the common internal tangent segment of two escribed circles:

$$\begin{aligned} N_1N_2 &= AN_1 - AN_2 = s - (s - b) = s - s + b \\ &= b. \end{aligned}$$

$\overline{NN_1}$ is the common external tangent segment of the inscribed circle and an escribed circle:

$$\begin{aligned} NN_1 &= AN_1 - AN = s - (s - a) = s - s + a \\ &= a. \end{aligned}$$

$\overline{NN_2}$ is the common internal tangent segment of the inscribed circle and an escribed circle:

$$\begin{aligned} NN_2 &= AB - BN - AN_2 = c - (s - b) - (s - b) \\ &= c - 2s + 2b = c - (a + b + c) + 2b \\ &= c - a - b - c + 2b \\ &= b - a. \end{aligned}$$

Thus, the lengths of all the tangent segments contained in the line that also contains c can be written in terms of a and b.

Extension

It is extremely important for students to be able to generalize their results. Therefore, the Extension should be done by all students. If necessary, discuss the first one or two relationships and have students do the others on their own. Point out that the relationships found previously could be found for the other tangent segments in the figure by similar methods. Thus, it is possible to generalize from the four equations.

For $N_1N_3 = a + b$: The length of the common external tangent segment of two escribed circles equals the sum of the lengths of the two sides that intersect it.

For $N_1N_2 = b$: The length of a common internal tangent segment of two escribed circles equals the length of the side opposite the vertex contained in the tangent segment.

For $NN_1 = a$: The length of the common external tangent segment of an inscribed and escribed circle equals the length of the side that intersects it.

For $NN_2 = b - a$: The length of a common internal tangent segment of an inscribed and escribed circle equals the difference between the lengths of the two sides *not* containing the tangent segment.

Ask students the lengths of other tangent segments in the figure. Have them describe the segment and then apply the appropriate relationship. For example, to find the length of $\overline{MM_1}$, students should first identify it as a common external tangent of an inscribed and escribed circle. They then know it equals the length of the side that intersects it, and $MM_1 = a$.

More Equicircles

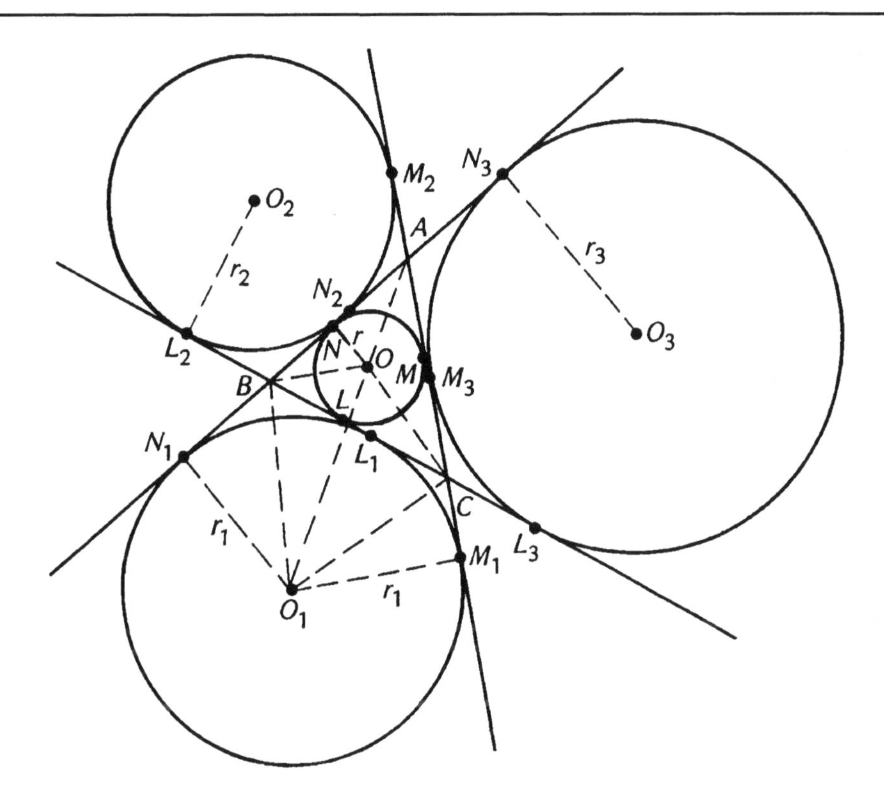

In "Equicircles" you found the lengths of various tangent segments in terms of a, b, and c, the lengths of the sides of $\triangle ABC$. There is also a relationship between the radii of the equicircles and the area of $\triangle ABC$.

First consider the radius of the inscribed circle and the triangles formed by $\overline{AO}$, $\overline{BO}$, and $\overline{CO}$. The area of $\triangle ABC$ is the sum of the areas of $\triangle ABO$, $\triangle ACO$, and $\triangle BCO$. If $\overline{AB}$ is the base of $\triangle ABO$, it has length c. What is the length of the altitude to $\overline{AB}$ in $\triangle ABO$? ____________

Why? ____

____

What is the area of $\triangle ABO$? ____________ What is the area of $\triangle ACO$? ____________ What is the area of $\triangle BCO$? ____________

Therefore, area $\triangle ABC =$ ____________.

What is r in terms of the area and semiperimeter of $\triangle ABC$? ____________

Now consider escribed circle O_1 and the triangles formed by $\overline{AO_1}$, $\overline{BO_1}$, and $\overline{CO_1}$:

$$
\begin{aligned}
\text{area}\,\triangle ABC &= \text{area}\,\triangle ABO_1 + \text{area}\,\triangle ACO_1 - \text{area}\,\triangle BCO_1 \\
&= \frac{1}{2}r_1c + \frac{1}{2}r_1b - \frac{1}{2}r_1a \\
&= r_1\left[\frac{1}{2}(c+b-a)\right] \\
&= r_1\left[\frac{1}{2}(c+b+a) - 2a\right] \\
&= r_1\left[\frac{1}{2}(a+b+c) - a\right] \\
&= \underline{\hspace{4cm}}\,.
\end{aligned}
$$

So $r_1 = \frac{\text{area}\,\triangle ABC}{s-a}$. That is, the radius of escribed circle O_1 equals the ratio of the area of $\triangle ABC$ to the difference between the semiperimeter and the length of the side to which O_1 is tangent. This ratio is true for any escribed circle. Therefore,

$r_2 =$ ________________ and $r_3 =$ ________________.

What is the product of the radii of the equicircles?

$r \cdot r_1 \cdot r_2 \cdot r_3 =$ _______________________.

Recall Heron's formula: area $\triangle ABC = \sqrt{s(s-a)(s-b)(s-c)}$. Thus,

$(\text{area}\,\triangle ABC)^2 =$ ____________________________________

and

$r \cdot r_1 \cdot r_2 \cdot r_3 =$ _______________________.

EXTENSION! The reciprocals of the radii of equicircles are also related. Use the values of r, r_1, r_2, and r_3 you already found to find this relationship.

Teacher's Notes for More Equicircles

Very few topics in high school geometry present as many different relationships as equicircles. For this reason, two activities on this topic are included in this book. "Equicircles" concentrates on the tangent segments formed. This activity explores relationships between radii of equicircles.

Although the topics presented here do not rely on anything developed in "Equicircles," "More Equicircles" does assume a knowledge of what equicircles are. In addition, students should be familiar with both the usual formula for the area of a triangle and Heron's (Hero's) formula.

NCTM Standards

1	2	3	4	5	6	7	8	9	10
•	•	•	•		•	•	•	•	•

Presenting the Activity

It may be necessary to briefly review the definition of equicircles given in "Equicircles" and to discuss how the figure on the student page is drawn.

By drawing $\overline{AO}$, $\overline{BO}$, and $\overline{CO}$, the area of $\triangle ABC$ is easy to find as the sum of the areas of the smaller triangles. Be sure students understand which sides have length a, b, and c. The length of the altitude to $\overline{AB}$ in $\triangle ABO$ is r, because the radius drawn to the point of contact of a tangent is perpendicular to the tangent. Whereas the length of $\overline{AB}$ is c, the area of $\triangle ABO = \frac{1}{2}rc$. Similarly, the area of $\triangle ACO = \frac{1}{2}rb$ and the area of $\triangle BCO = \frac{1}{2}ra$. Therefore,

$$\begin{aligned}\text{area}\,\triangle ABC &= \frac{1}{2}rc + \frac{1}{2}rb + \frac{1}{2}ra \\ &= r\left[\frac{1}{2}(a+b+c)\right] \\ &= rs\end{aligned}$$

and

$$r = \frac{\text{area}\,\triangle ABC}{s}.$$

Next, students find r_1 in terms of measures of $\triangle ABC$. $\overline{AO_1}$, $\overline{BO_1}$ and $\overline{CO_1}$ are drawn to form triangles. Each triangle has one side of $\triangle ABC$ as a base and r_1 as the altitude to that base. For example, in $\triangle ABO_1$, r_1 is the altitude to base $\overline{AB}$. Thus, the area of $\triangle ABO_1$ is $\frac{1}{2}cr_1$. The areas of triangles ACO_1 and BCO_1 are found in a similar manner. Then the area of $\triangle ABC$ is found in terms of the areas of these triangles. Notice in the fourth line of the equations, $2a$ is subtracted; thus, the expression will simplify to $r_1(s-a)$.

It's possible to show that $\frac{1}{2}(c+b-a) = s - a$ using lengths of segments as was done in "Equicircles." This method is shown next in case you have the inclination and the time to discuss it:

$$\begin{aligned} AN + AM &= (AB - BN) + (AC - CM) \\ &= (AB - BL) + (AC - CL) \\ &= (AB + AC) - (BL + CL) \\ &= c + b - a. \end{aligned}$$

However, $AN = AM$, so $AN + AN = c + b - a$ and $AN = \frac{1}{2}(c + b - a)$. Students discovered in "Equicircles" that $AN = s - a$, so $\frac{1}{2}(c + b - a) = s - a$.

Whereas the ratio described on the student page is true for any escribed circle,

$$r_2 = \frac{\text{area}\,\triangle ABC}{s - c} \quad \text{and} \quad r_3 = \frac{\text{area}\,\triangle ABC}{s - b}.$$

Now students find the product of the radii equals

$$\frac{\text{area}\,\triangle ABC}{s} \cdot \frac{\text{area}\,\triangle ABC}{s - a} \cdot \frac{\text{area}\,\triangle ABC}{s - c} \cdot \frac{\text{area}\,\triangle ABC}{s - b}$$

or

$$r \cdot r_1 \cdot r_2 \cdot r_3 = \frac{(\text{area}\,\triangle ABC)^4}{s(s - a)(s - b)(s - c)}.$$

This appears to be a very complex expression until Heron's formula for area is used. (*Note*: Some texts refer to this as Hero's formula. This is merely a difference in translation.) By the formula,

$$(\text{area}\,\triangle ABC)^2 = s(s - a)(s - b)(s - c)$$

and

$$r \cdot r_1 \cdot r_2 \cdot r_3 = \frac{(\text{area}\,\triangle ABC)^4}{(\text{area}\,\triangle ABC)^2} = (\text{area}\,\triangle ABC)^2.$$

Some students may wish to verify this relationship using specific values of a, b, and c. They can do this by using Heron's formula to find the area of $\triangle ABC$ and using the ratios on the student page to find r, r_1, r_2, and r_3. This relationship also holds under dimensional analysis, since both sides are of fourth degree.

Extension

Students should begin by finding the reciprocals of the radii:

$$\frac{1}{r} = \frac{s}{\text{area}\,\triangle ABC}, \qquad \frac{1}{r_1} = \frac{s - a}{\text{area}\,\triangle ABC},$$

$$\frac{1}{r_2} = \frac{s - c}{\text{area}\,\triangle ABC}, \qquad \frac{1}{r_3} = \frac{s - b}{\text{area}\,\triangle ABC}.$$

Allow them to experiment with these values for awhile. Then if they appear stymied, suggest they try adding the reciprocals of the radii of the escribed circles:

$$\begin{aligned}
\frac{1}{r_1}+\frac{1}{r_2}+\frac{1}{r_3} &= \frac{(s-a)+(s-c)+(s-b)}{\text{area}\,\triangle ABC} \\
&= \frac{3s-(a+b+c)}{\text{area}\,\triangle ABC} \\
&= \frac{s}{\text{area}\,\triangle ABC} \\
&= \frac{1}{r}.
\end{aligned}$$

Locus Methods

• A

x

How would you describe the set of all points in a plane at a distance x from point A in the preceding figure? ______________________________

____

Construct the set of points you described using the given value of x shown in the figure.

The set of *all* the points, and *only* those points, that satisfy a given condition is called a *locus*. In the foregoing problem, the locus of points x distance from A is a circle with center A and radius x.

Now suppose you are to construct a circle of given radius r, that passes through a given point M and is tangent to a given circle P with radius R. This problem requires finding two loci:

1. The locus of centers of circles of radius r that pass through M.
2. The locus of centers of circles of radius r that are tangent to circle P.

The *intersection* of the two loci is the center of the circle to be constructed.

Describe the locus of points in item 1: ______________________________
Using the following figure, construct this locus.

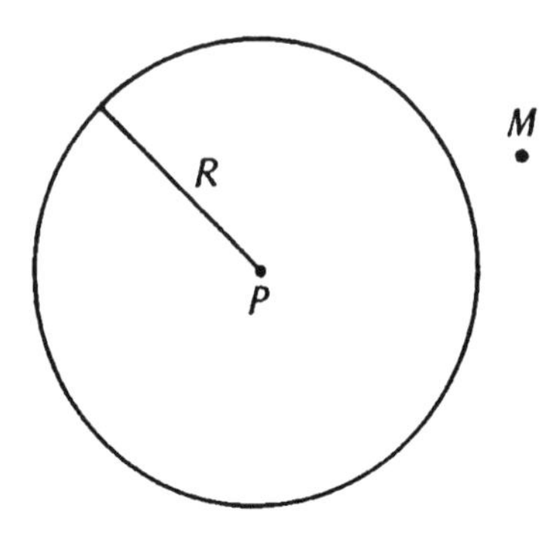

r

Now describe the locus of points in item 2: ______________________________
Construct this locus using the same figure.

How many points of intersection of the two loci are there? _______ For the same figure, how many solutions are there? That is, how many circles can be constructed that satisfy the given conditions? _______________________

The number of solutions to this problem depends on the relative positions of M and P and on the relative lengths of r and R. In the given figure, $r > R$ and $d > R$, where d is the distance from M to P. How does d compare to $R + 2r$? _____________ Suppose circle P and r remain as in the figure, but $d = R + 2r$.

How many solutions will there be? _______ Sketch this case.

How many solutions will there be if $d > R + 2r$? _______ Sketch this case.

If $d = R$, where is point M? _______________________________

How many solutions will there be in this case? _____________________

Where is point M if $d = 0$? _______________________________

What must be true about R and r for there to be a solution in this case? _____

____

EXTENSION! In the cases listed, $d > R$, $d = R$, or $d = 0$. Suppose circle P and r stay the same, but $d < R$. (M is in the interior of circle P.) Sketch the situations that will give (1) two solutions, (2) one solution, and (3) no solution. How does d compare to $R - 2r$ for each situation?

Teacher's Notes for Locus Methods

Locus methods are often the best way to approach certain types of problems and constructions. Analysis of a problem such as the one given in this activity is almost impossible for a high school student unless locus methods are used. This activity gives students a taste of how to analyze a mathematical problem so every possible case is considered—an extremely important area of more advanced mathematical proof and scientific experimentation.

NCTM Standards									
1	2	3	4	5	6	7	8	9	10
	•	•	•		•	•	•	•	•

Presenting the Activity

The first problem on the student page is primarily for assessment. Any student unable to complete this problem quickly and easily should *not* proceed with the activity. Emphasize that a locus must include *all* the points that satisfy the condition *and* every point on the locus must satisfy the condition.

Discuss the two loci for the problem. The first locus is a circle of radius r with center at M. Every circle with radius r that also passes through M will have its center on this circle. The second locus consists of two circles concentric with circle P, one with radius $R + r$ and one with radius $R - r$. Every circle with radius r that is also tangent to circle P will have its center on one of these two circles. The two loci are shown with dashed lines in the following figure.

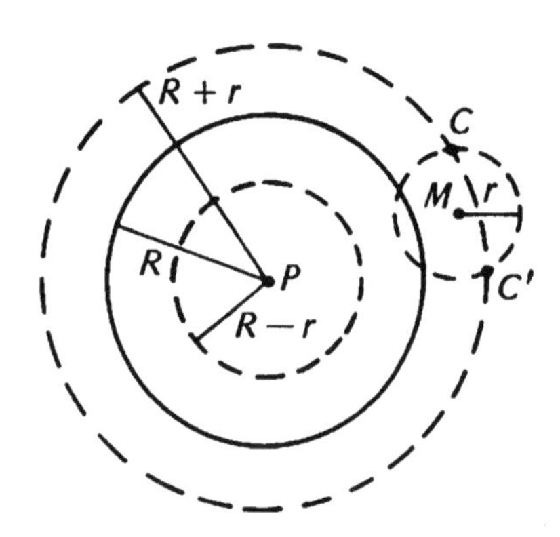

C and C' are the points where the loci intersect and are the centers of the two circles of radius r that pass through M and are tangent to circle P. Thus, there are two solutions for the given conditions. Now students are asked to analyze what happens when the position of M changes. This analysis uses d as the distance from M to P and considers how the solutions change as d changes. (Notice that radii R and r do not change.) The first cases considered are those where $d > R$. If $d < R + 2r$, we have the preceding situation. If $d = R + 2r$, there is one solution:

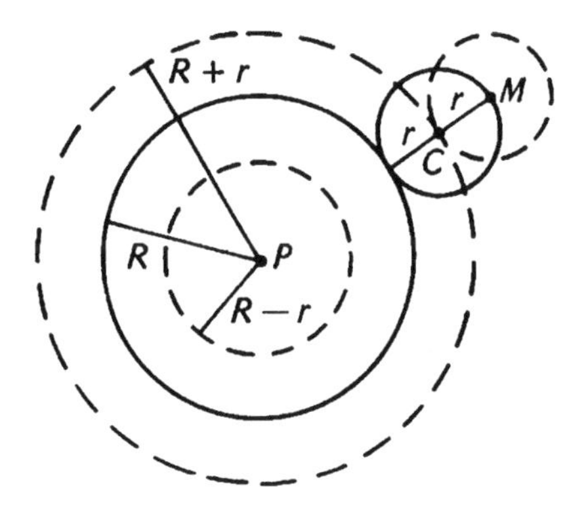

If $d > R + 2r$, there are no solutions:

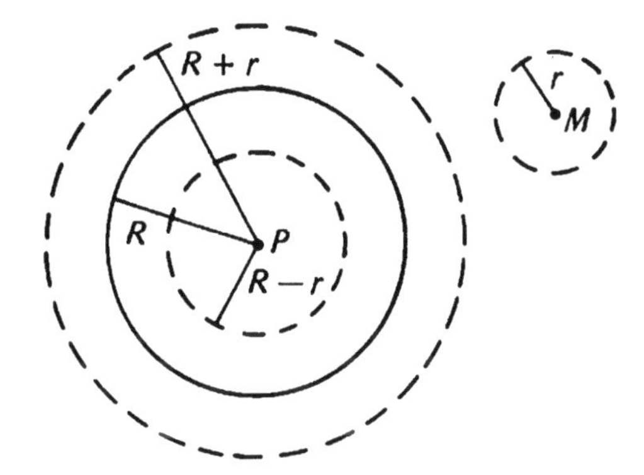

If $d = R$, M is on circle P and there are two solutions:

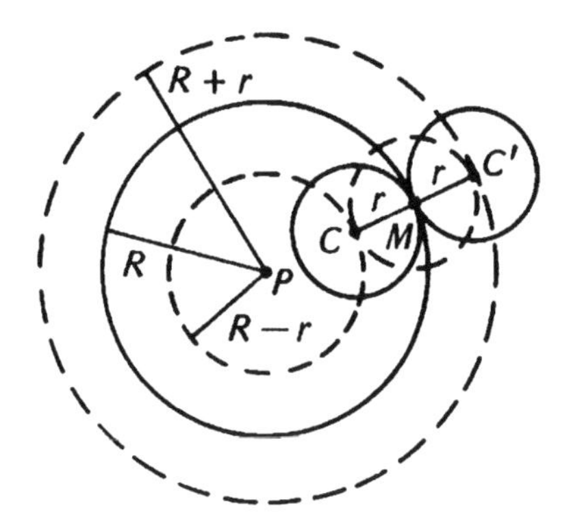

When $d = O$, M and P coincide and circles M and P are concentric. There is no solution unless $R = 2r$, and then there will be infinitely many solutions.

Extension

The Extension completes the analysis for $r < R$. In these three cases, $d < R$ and M is in the interior of circle P. If $d > R - 2r$, there are two solutions. If $d = R - 2r$, there is one solution. If $d > R - 2r$, there are no solutions. These three cases are analogous to the three cases for $d > R$, where the number of solutions was determined by the number of intersections of circles M and $R + r$. For $d < R$, the number of solutions is determined by the number of intersections of circles M and $R - r$.

More advanced students may want to consider the cases when $r \geqslant R$. For $r = R$, the circle of radius $R - r$ reduces to point P and the solutions are

$d > R$: $d < 3r$, two solutions,

$d = 3r$, one solution,

$d > 3r$, no solutions;

$d = R$: one solution;

$d < R$: no solutions.

For $r > R$, the circles for the second locus have radii $r + R$ and $r - R$. The following cases will result.

$d > R$: $d < 2r - R$, four solutions,

$d = 2r - R$, three solutions,

$d > 2r - R$: $d < 2r + R$, two solutions,

$d = 2r + R$, one solution,

$d > 2r + R$, no solutions;

$d = R$: two solutions;

$d < R$: no solutions.

Sketches of the figures will help with the analysis.

Resource A: List of Activities and the NCTM Standards Addressed by Each

Name of Activity	NCTM Standards									
	1	2	3	4	5	6	7	8	9	10
Constructing Segments		•	•	•		•	•		•	
Constructing Radical Lengths	•	•	•	•		•	•	•	•	•
Trisecting a Circle	•	•				•	•		•	•
Constructing a Pentagon	•	•	•	•		•	•		•	•
Trisecting an Angle	•	•	•	•		•	•		•	•
Constructing Triangles	•	•	•	•		•	•	•	•	•
The Pythagorean Theorem	•	•	•	•		•			•	•
The Golden Rectangle	•	•	•	•		•	•		•	•
The Golden Triangle	•	•	•	•		•	•	•		•
The Arbelos	•	•	•	•		•	•		•	•
Ptolemy's Theorem		•	•	•		•	•		•	•
Ceva's Theorem		•	•	•		•	•		•	•
Stewart's Theorem	•	•	•	•		•	•		•	•
Simson's Theorem		•	•	•		•	•		•	•
Napoleon's Theorem	•	•	•	•		•	•		•	
Taxicab Geometry	•	•	•	•		•			•	•
Transformational Geometry-Symmetry			•					•		
Projective Geometry			•				•		•	•
Spherical Geometry			•			•	•	•		
This Wraps It Up	•	•	•	•		•	•		•	•
Regular Polyhedra	•	•	•			•	•		•	•
Cavalieri's Principle	•	•	•	•		•	•	•	•	•
The Jolly Green Giant?	•	•	•	•		•	•	•	•	
Mathematics on a Billiard Table			•	•		•	•		•	•
Bypassing an Inaccessible Region			•	•		•	•			•
The Inaccessible Angle		•	•	•		•	•	•	•	•
Minimizing Distances		•	•	•		•	•	•	•	•
Problem Solving—A Reverse Strategy		•	•	•		•	•	•	•	•
Geometric Fallacies		•	•	•		•	•	•	•	•
The Nine-Point Circle		•	•	•		•	•	•	•	•
Equicircles	•	•	•	•		•	•		•	•
More Equicircles	•	•	•	•		•	•	•	•	•
Locus Methods		•	•	•		•	•	•	•	•

The Corwin Press logo—a raven striding across an open book—represents the happy union of courage and learning. We are a professional-level publisher of books and journals for K-12 educators, and we are committed to creating and providing resources that embody these qualities. Corwin's motto is "Success for All Learners."